Virtual Roaming Data Services and Seamless Technology Change – GSM, LTE, WiFi, Satellite, CDMA

Cover picture
Roaming in mid Atlantic, the submarine cruiser Surcouf hunting the Bismarck supply tankers, May 14-June 14, 1941.

RIVER PUBLISHERS SERIES IN COMMUNICATIONS

Consulting Series Editors

MARINA RUGGIERI
University of Roma "Tor Vergata"
Italy

HOMAYOUN NIKOOKAR
Delft University of Technology
The Netherlands

This series focuses on communications science and technology. This includes the theory and use of systems involving all terminals, computers, and information processors; wired and wireless networks; and network layouts, procontentsols, architectures, and implementations.

Furthermore, developments toward newmarket demands in systems, products, and technologies such as personal communications services, multimedia systems, enterprise networks, and optical communications systems.

• Wireless Communications
• Networks
• Security
• Antennas & Propagation
• Microwaves
• Software Defined Radio

For a list of other books in this series, visit
http://riverpublishers.com/river publisher/series.php?msg=Communications

Virtual Roaming Data Services and Seamless Technology Change

− GSM, LTE, WiFi, Satellite, CDMA

Arnaud Henry-Labordère
Professor Ecole Nationale des Ponts et Chaussées
Chairman Halys, France

LONDON AND NEW YORK

Published 2014 by River Publishers
River Publishers
Alsbjergvej 10, 9260 Gistrup, Denmark
www.riverpublishers.com

Distributed exclusively by Routledge
4 Park Square, Milton Park, Abingdon, Oxon OX14 4RN
605 Third Avenue, New York, NY 10017, USA

First issued in paperback 2023

Virtual Roaming Data Services and Seamless Technology Change – GSM, LTE, WiFi, Satellite, CDMA / by Arnaud Henry-Labordère.

© 2014 River Publishers. All rights reserved. No part of this publication may be reproduced, stored in a retrieval systems, or transmitted in any form or by any means, mechanical, photocopying, recording or otherwise, without prior written permission of the publishers.

Routledge is an imprint of the Taylor & Francis Group, an informa business

Publisher's Note
The publisher has gone to great lengths to ensure the quality of this reprint but points out that some imperfections in the original copies may be apparent.

While every effort is made to provide dependable information, the publisher, authors, and editors cannot be held responsible for any errors or omissions.

ISBN 13: 978-87-7022-974-6 (pbk)
ISBN 13: 978-87-93102-23-1 (hbk)
ISBN 13: 978-1-003-34001-0 (ebk)

Table of Contents

Introduction

The subject is "Virtual Roaming for data services" and "Seamless Technology change" also called "Number Continuity". Virtual Roaming for voice and SMS services was covered in a previous book [0.8]. What is the difference brought by "Virtual"? It means that it allows a subscriber to visit a network with which his home network does not have an agreement; this is "virtual roaming":

- Commercially, a "Third party" is used which has an agreement with both,
- Technically, the routing is not just at the network layer protocol (performed by a STP for MTP3 or M3UA /SS7, by a SCCP gateway for SCCP/SS7 or by an IP router for IP. The routing involves changing the *content of the protocol messages* at application level.

The "Seamless Technology change" allows a user to keep all his services including reception of calls and SMS sent to his usual number when he switches his GSM or CDMA handset to the WiFi mode or when he uses a satellite phone instead of his usual handset.

Instead of using the vague « Roaming Hub » term, one of the meaningless terms likes « SMS Gateway or such generalities, we use:

- SS7 Hub (for MAP and CAMEL),

as well as:

- GTP Hub (for the GPRS data protocol used in GSM 2.5G, 3G and LTE 4G
- RADIUS Hub (authentication and accounting)
- SIP Hubs and RCS Hubs (Rich Communication Suite, an extension of SIP Hubs)
- Diameter Hub (LTE virtual roaming hubs)
- MMS Hubs (MMS protocol using the SMTP « mail protocol » bearer).

which are the subjects developed in the following chapters. The main parameters used for routing between networks are:

- IMSI for GTP Hubs and MMS Hubs
- realm (more recent term than « domain name ») for RADIUS Hub and SIP Hub.

In addition, using an "Address Resolution service" such as ENUM or the SS7 interrogation of the HLRs, it is necessary to obtain the IMSI from the MSISDN for

- RCS Hubs
- MMS Hubs
- SMS Hubs (which is included in the most recent SS7 Hubs).

This is text book for a graduate course on mobile networks. The reader is assumed to have a good knowledge of SS7 as well as knowledge of virtual roaming. References are given if this was not the case.

Chapter 1 is an introductory complement and a 4G refresher used in the rest of the book. Other virtual roaming implementations not using "Roaming Hubs" but based on multi-IMSI HLR or HSS(LTE 4G) are explained as well as new automatic test simulators to simplify the expensive roaming relation tests of the MNOs(Mobile Network Operator) or the Roaming Hub providers.

Some key principles of the GRX network (the Internet reserved to the mobile operators) are explained. Including / the use of DNS(Domain Name Servers) in the GRX network as IP networks are used for signalling of mobile data services (2G, 3G and 4G) as it is essential to understand the DNS tree mechanism for mobile data network access. It is also essential to explain the hierarchy of the participants in the GRX, mobile operators, GRX providers and the way the BGP protocol is used to create dynamically the routing between the mobiles and the web sites. This is useful to understand the possibilities of redirecting the traffic for various purposes.

The path between a mobile and a category of web services (Internet, MMS, VPN, etc..) is directed by a profile setting in the handsets which is cumbersome to change by the user, this is why automatic OTA (Over The Air) provisioning is used. We had previously fully covered the subject in a 2009 book. It has become more complicated since new profile setting standards have become common (Android, Windows OS, Apple OS) in parallel with the OMA standard. A practical comparison is given. This is useful in understanding the interest of methods implementing the Local Break-Out without changing the handsets profiles of Chapter 3.

There is also a complement on Location Services for 4G which may use MME or AAA servers as the serving node. The LCS methods are very similar to those used in 2G and 3G.

Chapter 2 gives the general principles of Virtual Data Roaming, which are explained in all details in other specific chapters. They include the conversion Hubs between the different access protocols such as GSM<->SIP, GSM<->WiFi, GSM<->CDMA, GSM<->LTE. We explain the role of the DIAMETER and RADIUS protocols (although RADIUS is not used in 4G but it remains as a legacy of the previous authentication method even in 3G.

We give as Table 1 the description of the content of this book. OA is « originator », DA is « destination ».In the last column, we have indicated the incomplete solution which could be used alternatively to a « Hub » in certain cases.

Table 1 Topics of the book

Type of Hub	*Application*	*Main Routing Key*	*Main Routing Key obtained by "address resolution" using*	*Type of equipment for alternative low level routing solution*
GTP Hub	GSM and LTE virtual data roaming	IMSI		IP router for the GTP Hub ENTRY
MMS Hub	MMS interworking	« Realm MMS » based on IMSI	MSISDN of DA	None
RADIUS Hub	Wifi international roaming	Realm of OA		DNS
SIP Hub (uses also RADIUS Hub for IP access)	SIP VoiIP and message interconnection	Realm of DA		DNS
RCS Hub	RCS interconnection	« Realm IMS » of DA	MSISDN of DA	None
Diameter Hub	LTE international roaming	Realm of DA		DNS

Table 2 Topics excluded (There are some topics that are not covered in the book)

Type of Hub	Application	Main Routing Key	Main Routing Key obtained by "address resolution" using:	Type of equipment for alternative low level routing solution
SS7 Hub (SMS) see [0.7]	SMS interworking	IMSI	MSISDN of DA	STP (MTP level routing) or SCCP Gateway (SCCP level routing)
SS7 Hub (Voice) see [0.8]	Virtual Roaming Hub	IMSI	IMSI of DA	STP (MTP level routing) or SCCP Gateway (SCCP level routing)

To properly handle the Mobile Number Portability (MNP) for SMS interworking of countries which have implemented it, the use of a SS7 Hub with a search algorithm is mandatory. Take the case of an operator which uses multiple SMS Hub providers from a given MNP country. A statistic allocation of the NDCs among these Hubs to terminate the SMS is a defective approach, as it does not handle properly the « ported-in » and « ported-out » case.

Mobile Number Portability is now a common service allowing users to keep their number if they change their subscribed operator. It was first deployed in Hong-Kong (1999). A full explanation of the various classical technical solution is included, as well as a solution which does not need a centrally created data base while still providing the "direct routing".

« Number Continuity » corresponds to the same service but with the *switch to another technology*, because the subscribed main terminal does not have coverage. The new terminal may be a PC, a smartphone / WiFi, a "satphone", a GSM phone if the main one is CDMA, with (almost) the same service transparently. Making calls or sending SMS with its normal CLI shown, receiving calls, SMS, MMS to his normal GSM number (unlike « Skype »). The implementation uses virtual data roaming systems, it is covered in different chapters: WiFi, satphone, CDMA->GSM.

Chapter 2 is then a general short technical introduction to the various detailed chapters which will be sufficient for marketing specialists as they need to understand the issues without a need to deeply understand the implementation. The other chapters may be read fairly independently.

Chapter 3 is a main chapter of the book. It includes an explanation of the System Architecture Evolution (SAE) which defines the 4G evolved packet core. As the main focus is GTP hubs, it gives a very detailed explanation of the P-GW with call flows and traces, as a 3G or 4G GTP Hub is based on the combination of a GGSN-P-GW (most modern implementations handle both the 3G and the 4G packet cores) and a SGSN or MME.

There is also the recent discussion on the obligation to offer the "Local Break Out" for data in Europe. The BEREC (Body of European Regulators for Electronic Communications) has prepared for the European Commission a new regulation applicable on 1rst July 2014 which will allow (this is called "decoupling") the separate sale of roaming services. This will allow "alternative roaming providers"(ARP) to provide roaming services in order to allow cheaper use of services while roaming. This concerns only data services using single IMSI (dual IMSI is considered as not sufficiently practical for the target market), and not voice and data services also considered not practical by BEREC. The LBO model consists of local provision of retail data services *by the visited network* without any intervention of the HPLMN except for the SIM card authentication. In BEREC's model, after purchasing the LBO service from the visited network, the roaming visitors would provision a common APN "EUinternet" (this the scheme also proposed for LTE Local Break-Out). The ARP could eventually facilitate this provisioning of the RP with an OTA server (difficult: Iphone and Blackberry devices do not follow the standard OMA OTA setting), but then the original setting would have to be restored when the visitor is back home. Technically the VPLMN SGSNs may be configured to accept "EUinternet" independently of the HLR profile although this APN may be systematically included in the profile. There is no need for a special implementation and a standard GGSN or PDN Gateway (4G) of the VPLMN will provide the local access to Internet, as the subscriber is connecting *to* the visited GGSN which docs not need to have the GTP Hub function.

However the method requires that the visitor has set a new APN for internet and is able to *have his telephone internet setting changed to this APN* "EUinternet". This is *a major practical constraint* which make BEREC's scheme rather limited in scope in our opinion. Chapter 3 describes a better alternative and original solution based on a GTP Hub included in the VPLMN GGSN, which goes beyond the requirements, as the new APN "EUinternet" provisioning phase is not needed, and the LBO service may be modulated as such: the entire Internet, only for a list of web services, only for certain services (VoIP, etc..), while other services such as

MMS still use the data connection with the HPLMN. The visitors with LBO are simply provisioned with their MSISDN in the standard RADIUS server with a data volume credit. With the default LBO setup for all visitors for the VPLMN WEB service, there is an easy way for self LBO registration with a credit card.

Chapter 4 explains the implementation of WiFi roaming using RADIUS Hubs and recent charging solutions for GSM and LTE using Policy Charging and Control systems, which are a fundamental brick of the recent 3G networks and mandatory in 4G. It allows to offer a Quality of Service (variable bandwidth) customized to each user and to the particular service needed (VoIP needs a fast transmission which email does not).

Chapter 5 makes use of these Hubs for VoIP services and the recently introduced Rich Communication suite (RCS) which interconnection rules are in the process of being implemented.

Chapter 6 is about the virtual roaming implementation of 4G(LTE) networks, which have replaced the traditional MAP/SS7 protocol by S6a / DIAMETER. It is common throughout the world, unlike the difference which exist between MAP GSM and MAP IS-41 (CDMA).

Chapter 7 is for MMS Hubs. This is entirely new compared with a chapter in my [0.7); it describes the practical implementation of a MMS Hub: configuration of the DNS, multiple destination sending, address resolution (HLR interrogation using SS7 as ENUM is not used) to handle the Mobile Number Portability, and the fast shortest path algorithm which simplifies the provisioning.

Chapter 8 is a full explanation of WiFi<->GSM number continuity including the authentication of subscribers using the EAP-SIM in the Access Points and a RADIUS server. We explain the use of OTA SIM to set the order of preferences of the WiFi connections.

Chapters 9 to 10 are « transversal » and deal with the number continuity: between GSM <-> IP and GSM <-> Satellite. This chapter gives a theory for the mathematical modelling of the «Steering of roaming», an opportunity to find another practical application of Operations Research.

In chapter 9, we give the orbit and drift computations to give an understanding of the station keeping including Telecom satellites as I did not found it covered with enough completeness anywhere else: how to compute the maximum daily East-West drift of a geostationary satellite gets a practical answer.

Chapter 10 explains the GSM<->CDMA number continuity, in particular applied to the CDMA based satellite handsets of Chapter 9. It may look antiquated to speak about the CDMA protocol in 2013 when focus is on 4G LTE. However the purpose is much broader that you may think, as > 10% of worldwide terrestrial mobiles are not GSM and CDMA is rather important on certain continents. The interworking, in particular for SMS, was never properly implemented using SS7, which is much better (transparency), until 2013 (other GSM<->CDMA SMS interworking use a very crude IP connection between SMSC and « SMS Hubs » which does not relay whether the SMS really arrived. The presentation is both simplified and more detailed than in my previous 2004 and 2009 books aiming at providing a reference for developers and telecom history students, using a detailed comparison based on a rather wide knowledge of the GSM protocols nowadays. To be sure, detailed structured CDMA traces and explanations are given and a fair understanding of CDMA/IS-41 can be achieved quickly. I have observed after more than 40 years of teaching that comparisons with other experiences are the foundations of lasting knowledge.

Steering of Roaming (SoR) has been implemented for many years. Chapter 11 gives a most efficient anti-steering implementation to contribute to the armor-big gun battle. It is amusing to hear experts saying anti-steering is illegal, when in fact the most popular method of SS7 steering is an abuse by the major operators as it creates extraneous traffic without any revenue for the rejected MNOs. The steering performed by preferences in the SIM card is not objectionable.

Chapter 11 explains a method using "Roaming Hubs" to provide the anti "steering of roaming" capability. To be comprehensive, I should also have explained anti-antisteering solutions, but it should be enough to provide the understanding and a contribution, in the telecom area, to the armor-big gun battle.

Chapter 12, uses elementary Markov chains probabilities to compute the efficiency of SoR and of Gateway Location Registers(GLR) to improve the roaming revenues. This is not something which was previously analytically covered.

Chapter 13 concerns Location methods (LCS), with a full update concerning LTE. The call flows are accurate for someone developing or trouble shooting an eSMLC. Special emphasis has been given to the U-TDOA method with LMU as this is definitely (including 3G) the best method to handle the emergency services of legacy phones. I have enclosed the mathematical considerations which are lacking in most commercial presentations with measured processing times of the mathematically rigorous Resultant-Sturm. The possibility to trigger alerts for Mobile Advertisement services is shown to be possible in standard LTE networks.

A very hot topic with the spread of LTE is the "Policy Charging and Control", in particular the allocation of priorities for the data services based on the subscriber profiles. Chapter 14 describes PCC implementations of the GGSN or PDN Gateways with "preemptive priorities" able to offload dynamically the lesser priority users.

A very comprehensive and up-to-date Acronym list is provided at the end. For the figures, there are 2 types:

-   message flow figures such as Figure 2.1 to show the transforms accomplished by a Hub
-   system diagrams such as Figure 2.3 (Rich Communication Suite interco)

This is a text book, with many new results for students and professionals and includes many real case traces in the body of the explanations; this is the preferred method when telecom subjects are taught. My previous Operations Research students (my teaching specialty for over 30 years) will find that simple mathematical models help to find some useful quantitative results in certain chapters.

My deep thanks to the publisher's team, to the members of the Halys R&D development team, especially Benoit Mathian, Waël Manaï, Sébastien Cruaux, Gilles Duporche, to Pascal Adjamagbo(IMJ) for his help in Chapter 9, to Laurent Gaignerot and to Oxana Aufort for her preparation of the text.

BY THE SAME AUTHOR

Mathematics

[0.1] « Méthodes et Modèles de la Recherche Opérationnelle », Vol.3 (with Arnold Kaufmann), Dunod 1973, translated to English « Integer and Mixed Programming », Addison Wesley (1976), Russian(MIR, 1975), Spanish (CCSA, 1975), Romanian(1976).

[0.2] « Exercices et Problèmes de Recherche Opérationnelle », Masson, 1976.

[0.3] « Analyse de données », Masson, 1976.

[0.4] « Recherche Opérationnelle », Presses des Ponts et Chaussées, 1981.

[0.5] « Cours de Recherche Opérationnelle», Presses des Ponts et Chaussées, 1995, Vol.1, Linear and nonlinear programming, Graph theory,

[0.6] « Cours de Recherche Opérationnelle», Presses des Ponts et Chaussées, 1995, Vol.2, Optimal control and infinite dimension optimisation, Game theory.

Telecommunications

[0.7] « SMS and MMS interworking in Mobile Networks », (with Vincent Jonack), Artech Publishing House, 2004.

[0.8] « Virtual Roaming Systems for GSM, GPRS and UMTS », Wiley, 2009.

1

Complements on Roaming Architectures for Voice – SMS Services and Virtual Data Roaming

Little John, my own dear friend, mark, I pry thee, where this arrow lodges,
there my grave be digged, and let my weary bones be not disturbed.
Drawing the bowstring to his ear, he sped the arrow out of the open
window.
−The Merry Adventures of Robin Hood, Edward Pyle (1853–1911)

We assume that the reader is already familiar with virtual roaming solutions based on "Roaming Hubs" [1.1] and will explain here the specific businesses which have other solutions than roaming hubs. These Roaming Hubs were designed for small MNOs wishing to extend the roaming reach of their outbound subscribers by using one or several "auxiliary IMSIs" provided by large MNOs acting as "sponsors". A "SIM Tool Kit" in the SIM card chooses automatically the IMSI to use in a given visited country, including the optimization of cost. The service is provided by a "Third party" which provides the Roaming Hub and the financial clearing acting as a single contact for the technical and commercial relation. Since a few years, HLR vendors (and now HSS for LTE) have developed solutions which address different markets, in particular, Critically Important Services or the migrant market. We will not discuss the trivial case of systems which require a manual operation (change of SIM card, manual selection of IMSI when there is a multi-IMSI SIM card), only architectures where there is either a full number continuity (the transparent case) or at least where the subscriber can be called on any MSISDN associated with any of the IMSIs.

We also provide here the necessary background on DNS to understand how the hosting of virtual GGSN (3G) or PDN Gateways (4G) is possible and also how a Local Break Out for the data service can be setup.

An important issue associated with Roaming Hubs has been the business but also the interco test simplification. The architecture of a "simulator and test platform" is presented. It performs the main tests without needing to exchange physical SIM cards.

1.1 Multi-IMSI Solutions for Virtual Roaming: Roaming Hub-Based and Multi-IMSI HLR-Based

There are three very different business cases:

- the well know case [1.1] of "carriers" such as Belgacom, Comfone, France Telecom, Telecom Italia, Telefonica, Telecom North America, etc., which offer to MNOs the possibility to extend the roaming coverage of their outbound subscribers. They use a roaming hub, and have no HLR to register individual subscribers.
- The MVNOs (the Mobile Virtual Network Operators, their own radio coverage is not necessary), selling multiple numbers to individual subscribers so that they have better rates in the country where they reside mostly and their own foreign country which they are most likely to visit every year (migrants case). *The users have several numbers.* Cases of White Mobile Label (Liechtenstein) include foreign workers in Europe, etc... In this case the MVNO needs its own "multi-IMSI" HLR as explained below, such as provided by several vendors (Huawei, Tekelec, etc.).
- The MVNOs for Critical Service administrations. They have a single number with the possibility of national and international roaming with a sponsor's IMSI in their dual-IMSI SIM card.

The traditional roaming hub case has the advantage that it can also handle the first case and also the third with a standard HLR. In the second, the MVNO must have its own distribution channel to sell individual multi-IMSI cards. He must also negotiate with "sponsors" to allocate IMSIs (and MSIDNs numbers).

1.1.1 Multi-IMSI Solutions: Roaming Hub-Based

The principle is reminded in the Figure 1.1. The main characteristic is that even when the mobile is using an auxiliary IMSI from one of the sponsors, the MSISDN which is *allocated is the MSISDN nominal from his home operator.* The Roaming Hub gets it from the HLR and passes it to the VLR. This way when a call or a SMS is sent, the called party sees the original

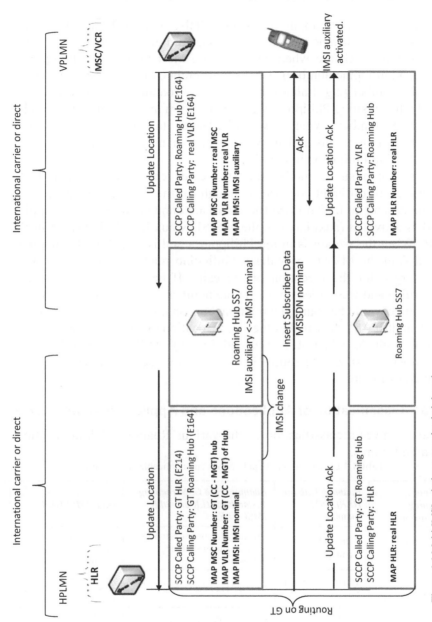

Figure 1.1 Multi-IMSI solutions: roaming hub–based.

1.1.2   Multi-IMSI Solution: Multi-IMSI HLR-Based

Take the case of a Philippino working in Italy. He must be able to receive calls at no cost when he works in Italy (thus an Italian number), but also to be called at no charge when he is visiting his family in the Philippines (hence, a Philippine number). This is a different case of the first. If he had a unique number (e.g. Italian), he would pay roaming charges when he is called while visiting Philippines. It is better for him to be called then on his Philippines number. Typically when he leaves Italy, he sets unconditional call forwarding to his Italian VMS which a message "please call my +6387123456 Philippino number".

To have a single model, all SIM cards could have several IMSI auxiliaries for different countries although for the case, a Philippino IMSI (auxiliary) and an Italian IMSI (nominal) are enough. When the Philippino IMSI is activated (auxiliary#2), the MSISDN auxiliary#2 is loaded in the VLR. When the worker is in the Philippines and is called on this number, the HLR of the MVNO provides a Philippino roaming number and he is not charged for the reception of the calls. If he had not done the call forwarding and was called on his Italian number, he would also receive the call, but he would be charged a roaming cost on his Italian subscription.

A multi-IMSI HLR makes the single location (MSC/VLR Global Title) record available when it is acceded by any MSISDN or IMSI of a subscriber. It has several entries in the AUC for each IMSI. The call flow is given by Figure 1.2

1.1.3   Classification of Applications and Applicable Architectures

Table 1.1 gives a classification of the various Roaming Hub applications and architectures.

Table1.1 Classification of applications and applicable architectures

Virtual roaming Architecture	Roaming Hub (no HLR) GSM roaming	Roaming Hub GSM roaming + standard HLR or multi-IMSI HLR	multi-IMSI HLR
Application Types	Outbound roaming	Critically Important Services (official administrations, defense)	Migrants' market (two countries mostly) or business travelers
Number of MSISDNs	MSISDN nominal Always	MSISDN nominal always	MSISDN changes depending on selected IMSI
Number of IMSIs	Several IMSI Auxiliaries	Several IMSI auxiliaries	Several IMSI Auxiliaries
Number of SIM cards for the user	One (with SIM Took Kit)	One (with SIM Took Kit)	One (with SIM Took Kit)

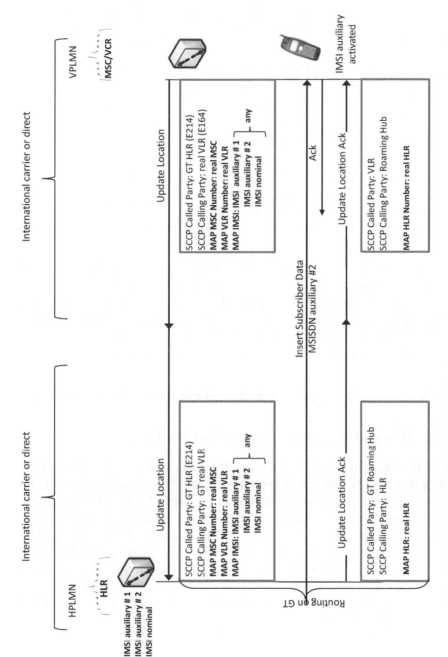

Figure 1.2 Multi-IMSI solution: multi-IMSI HLR-based.

Table 1.2 Classification of architectures for seamless technology change

Virtual roaming architecture	*Roaming Hub (no HLR) Satphone<-> GSM (Chapter 10)*	*MSC/VLR/SSF + SIP Gateway roaming (Chapter 8)*
Application Types	Outbound roaming in areas without GSM coverage	Outbound roaming using IP to save costs
Number of MSISDNs	MSISDN nominal Always	MSISDN nominal Always
Number of IMSIs	IMSI nominal+ One IMSI or MIN(CDMA) auxiliary	IMSI nominal
Number of SIM cards or MIN for the user	Two (standards), one GSM, one satphone	One (standard)

It is convenient to show also using Table 1.2, the various cases of « seamless technology change» covered in chapters 8 and 9.

To prepare for chapters on number continuity in particular with satellite phones, we show also the case of this service technology change where a satphone replaces the GSM. There are two different handsets and SIM cards then (or MIN in the handset if the satphone is CDMA-based).

1.2 Automatic Simulator and Test Platform for Roaming Tests

1.2.1   Utility of an Automatic and Test Platform

The test platform is used by MNOs or Roaming Hub providers to simplify and automate the interco tests. It suppress the traditional need for exchanging SIM cards and replaces this by "virtual SIM cards", that is just the IMSI numbers of "test cards" which are provisioned in the HLRs of the two partners (or just in the HLRs of the Roaming Hub's client). This provides great administrative and logistic savings (reliable mail post does not always exist in certain countries) and eliminates the risk of a fraudulent usage of physical test SIM cards for personal usage.

The tests which can be performed by such a test platform, without physical SIM cards, are:

- the standard IREG test: UL, SMS-MO, voice calls between two test cards,
- the Optional USSD and Supplementary Services testing,
- the CAMEL pre-paid test if a CAMEL agreement need to be tested,
- the SMS interworking test if the AA19 agreement is signed: sending SMS to the partner's numbers,
- the GPRS data roaming test,
- The MMS interworking test, if the agreement exists,
- The RCS(Rich Communication Suite) interco test, if the agreement exists.

With the platform, the tests are more comprehensive than with the standard manual tests. A full network test of the reachability of all the MNO's MSC/VLRs (their GTs are entered in the platform) and GGSNs associated with all the APNs (they are entered in the platform) is included, as well as the reachability of the HLR, of the SMSC and of the MMSC for the virtual SIM test cards.

The voice call test is also possible with softphones included in the test platform without the physical SIM cards (outgoing call and call reception).

Once configured, the tests are automatic and can thus be used for a regular check of the roaming partners. A test scenario can be created grouping all the RPs.

The simulator and test platform saves personnel and contributes to a better control of the roaming quality.

A test setup requires each partner to have at least two IMSIs (virtual SIM cards) of the other. For a Roaming Hub setup, only the provider needs the virtual SIM card and performs all the required testing for his client.

1.2.2 Simulator and Test Platform Principle

The two partners communicate through the SS7 network and the GRX network if data roaming is tested. The Test platform combines a MSC/VLR/SSF, a SGSN, a SMSC and a MMSC.

The configuration in Figure 1.3 only allows to test transactions sent by the HPLMN to his outbound subscribers (FWD_SM_MT, etc.) or to the HPLMN of the inbound visitors (UL, UL_GPRS, FWD_SM_MO for Map, Create PDP Context for GPRS, etc.). To test the other direction (transactions send TO the HPLMN), a test service with a second test platform is established.

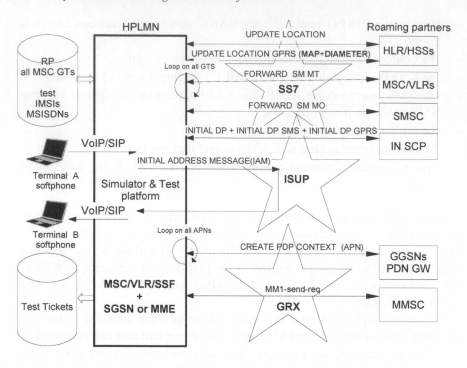

Figure 1.3 Automatic simulator and test platform principle for the outbound traffic.

For the outbound traffic, the test platform should include the GTs of all the MSC/VLRs of the tested MNO as well as its list of APNs to test that all the data services work properly for an outbound subscriber of the MNO. The inbound traffic simulator will *test only a small part of the SS7 and GRX networks*, but as it can simulate all the Roaming Partners' GTs, it will allow to determine errors in the setup of the MNO for particular RPs (SMS interworking, CAMEL agreements not properly configured, etc.). Such a pair of test platforms is much simpler than using a full traffic analyzer on all the international links and on the GRX interface.

A clever idea to allow the full testing of the outbound subscribers' traffic is to install "remote SIM card modems" in the main visited countries, as illustrated by Figure 1.4. *The SIM cards are in a central server and are connected to all the modems.* The testing service can then be provided by a service company for the testing of all the roaming agreements of many MNOs. The exchange of SIM cards can be suppressed. Only 240 modems would need to be installed to cover all the countries having mobile traffic.

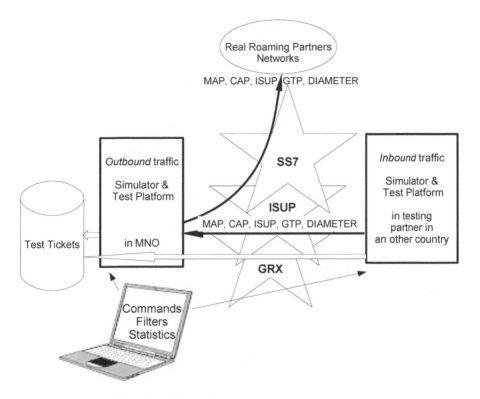

Figure 1.4 Testing the incoming traffic.

1.3 Role of the Domain Name Servers (DNS) for the Hosting of Virtual GGSN(3G) or PDN Gateways(4G)

For the exchange of data between networks, the mobile operators have the GRX network which is a secured IP network reserved for the mobile networks. A DNS provides an IP address when interrogated with a domain name derived from the IMSI ("Operator Identifier").There is a hierarchy of DNS (see Figure 1.3) which allows to provide a "Virtual GGSN" to mobile operators who do not have one. It is the same thing for 4G where MME is used instead of SGSN and "PDN Gateway" instead of GGSN. The scarcity of IP V4 addresses on the GRX network makes this quite attractive also for new mobile entrants which will find difficult to get one for their own GGSN (IPV6 are readily available but the different equipments on the GRX network are not all ready for it).

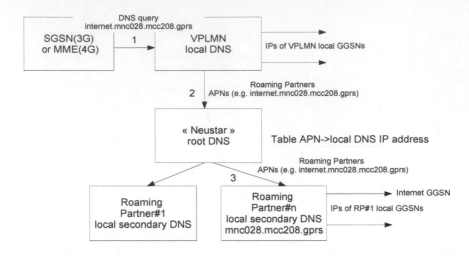

Figure 1.5 DNS Tree for GRX addressing.

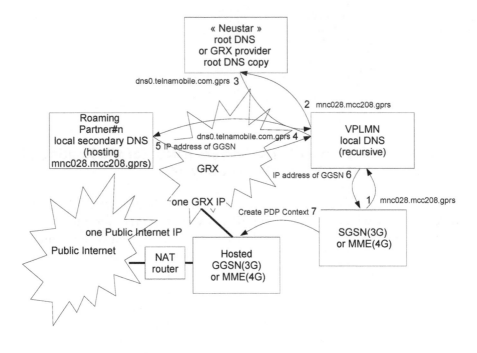

Figure 1.6 Hosting of virtual GGSN.

Figure 1.5 provides the tree structure for resolving domains on the GRX network. Local equipment (SGSN or MME) queries (1) their local DNS which contain (in general) only their own APNs. If the domain name is not local, the DNS queries (2) the "root DNS". With the domain name and a Table of APN->local DNS IP address, the root DNS queries (3) the concerned DNS which responds with the IP address of the Gateway GGSN for the user's service.

The five entities involved in setting up a virtual GGSN are:

A MVNO or Full MVNO (they have their own IMSI and HLR) **A** which uses a hosted GGSN of supplier **C**.

A visited VPLMN by a subscriber of **A** which has its own local DNS which in most cases has only the resolution for its own subscribers, except (Chapter 3) for the "Local Break Out" configuration if he wishes to provide the service.

A root DNS operated by Neustar on behalf of the GSM association. But the GRX operator of the VPLMN has a DNS root replicated from the Neustar root DNS and this is the DNS used in reality. In Figure 1.6, you can replace "Neustar" by "GRX provider of VPLMN".

An MNO **B** which sponsors **A** and includes **A**'s MCC-MNC and APN Operator Identifier in his IR21, the GSMa official document with the addressing information.

A supplier **C** of the hosted GGSN is connected to the GRX and can be shared among several operators such as **A**. It does have a DNS.

To explain with Figure 3.1, **A**'s subscriber activates a connection while visiting **B**. The SGSN uses the IMSI to create a domain name:

- Example 1: IMSI 208281234567890
                        -> mnc028.mcc208.gprs for 3G
- Example 2: IMSI 208281234567890
                            -> epc.mnc028.mcc208.3gppnetwork.org for 4G

The SGSN makes then a DNS enquiry(1) to its local DNS. The DNS does not find the domain in general and then chains the enquiry to the root DNS (2) which must have the table

mnc028.mcc208.gprs -> DNS IP of sponsor B (the enquiry can also be sent to the DNS of his GRX provider which is regularly copied from the root DNS).

The root DBS *returns (3) the address of the" local secondary DNS" (Figure 1.6 left)* to the local DNS of the VPLMN. As this is a recursive DNS, it will query (4) this "local secondary DNS" which returns successfully the IP address of the target GGSN or ME at the bottom of

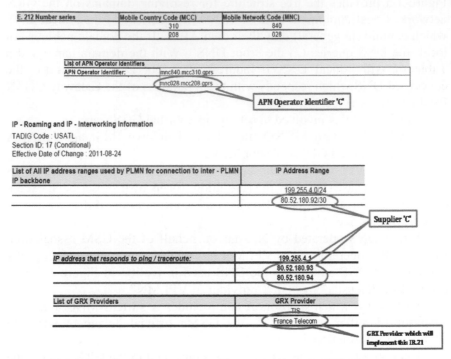

E. 212 Number series	Mobile Country Code (MCC)	Mobile Network Code (MNC)
	310	840
	208	028

List of APN Operator Identifiers	
APN Operator Identifier:	mnc840.mcc310.gprs
	mnc028.mcc208.gprs

APN Operator Identifier 'C'

IP - Roaming and IP - Interworking Information
TADIG Code : USATL
Section ID: 17 (Conditional)
Effective Date of Change : 2011-08-24

List of All IP address ranges used by PLMN for connection to inter - PLMN IP backbone	IP Address Range
	199.255.4.0/24
	80.52.180.92/30

Supplier 'C'

IP address that responds to ping / traceroute:	199.255.4.1
	80.52.180.93
	80.52.180.94

List of GRX Providers	GRX Provider
	TIS
	France Telecom

GRX Provider which will implement this IR.21

Figure 1.7 IR 21 of the DNS sponsor B (example Telecom North America TELNA).

Figure 1.6.

As a result, the SGSN will establish a session (7) with this GGSN (Create PDP Context of the GTP protocol).

This is simple technically; administratively, it is more complicated as explained in Chapter 3 because of the GSMa rules.
In order to have this implemented in the GRX, the DNS sponsor B (example TELNA) and the virtual GGSN supplier C will prepare the IR 21 of sponsor B with the agreement of the beneficiary **A**. And supplier **C** will ask his GRX provider (example: France Telecom) to implement in the root DNS. See such IR21 in Figure 1.7

1.4 Using a Single IP Address for the Public Internet: "Natting"

Public IP addresses are a scarce resource, almost exhausted for IPV4. Hence, GGSNs may use a single external public address for all the addresses, up to 65535 (a port number is on two octets) of the stations

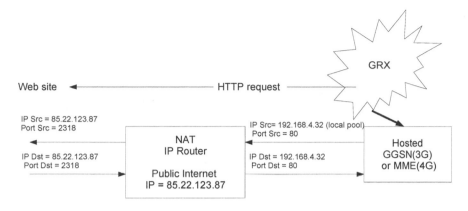

Figure 1.8 "Natting" of MS or UE IP addresses.

(local IP addresses "not routable"). The access to the public internet are through a "NAT IP router" in Figure 1.8, which replaces the local IP Source address and the Source Port by the single public address + a random Source Port (2318 in the example), as detailed in Figure 1.8. The router creates a table to do the inverse translation when it receives an answer. This "natting" mechanism is valid for all IP protocols, UDP (not connected), TCP (connected), SCTP (connected), etc. As a result, the supplier C needs only one IP address for his "GRX leg" (two with backup) and one public IP address (two for backup) for its "Internet leg" in Figure 1.4.

1.5 Structure and Hierarchy of the Main GRX Network for the SS7 Specialists

1.5.1 Difference between the Standard IP Routing in a Standard Server and the BGP Routing, and Comparison with SS7 SCCP Static Routing

A server may have several physical Ethernet ports to communicate with different LAN or WAN. The standard routing is static and uses "masks" for stating routing of a range of addresses.

192.168/16   ->Port 0
82.123.34.92/30 ->Port 1
52.12.14.94/29   ->Port 2

Table 1.3 Comparison of a SS7 GMSC and GRX BGP router

	SS7 SSCP Gateway (Hub of the SS7 network)	IP BGP router (Hub of the GRX network)
Routing to an adjacent peer	Static (primary/secondary Point code)	Dynamic (more than 2 adjacent peers possible), depending on the route evaluations received in BGP messages from the peers.
Routing to a non-adjacent peer through an adjacent peer	With SS7 MTP_ROUTE routing for MTP3(SS7) and M2PA(SIGTRAN). Can send a message to a distant Point Code which the adjacent peer has a route with.	No, for the IP main traffic. But the BGP messages may be broadcasted to distant AS through the IP route with an adjacent peer.
Information used to decide which route	SCCP Called Party GT	IP address of destination
Number of routes	Up to two (primary an secondary)	1 to N
Route tracing facility	No	Yes: "traceroute" utility uses the ICMP "Echo request" IP message.

The BGP routing is dynamic and computes a shortest path which also accounts for unavailability of routes. The routing changes depending on the BGP messages "advertise" received from the adjacent peers to which a BGP router is connected.

For those familiar with SS7, one can compare a SCCP Gateway and a BGP router in Table 1.3 above.

The physical network is quite similar. The IP BGP router has a physical link with each GRX adjacent partner. In the SS7 networks, a SCCP gateway has physical links with each IGP partner. However, SS7 is more flexible as "MTP3 routes" have the possibly to relay messages transparently to a distant partner (addressed by Point Code) through an adjacent route. Such a possibility does not exist in IP networks.

1.5.2 Example of Connection and Routing in the IP Networks GRX and Internet

1.5.2.1 Connection of the MNOs to the GRX

The MNOs have "IP Gateways" (called GGSN or PDN Gateway in LTE) which connect simultaneously to the GRX IP network to exchange signaling with their roaming partners, and to the Internet to allow their subscribers to access the various web services. Their firewall is an example of equipment which is connected to the two networks with different IP addresses for each.

In Figure 1.9, you see on the left an MNO which is connected to a single GRX provider. The MNO AS (Autonomous System) may use a private ASN (AS Number = 64912) as the routing policy with his GRX provider (ASN = 35030) is not visible from the other AS and this AS does not transit any IP traffic from other networks. A private ASN may be obtained from APNIC or a NIR, in this case if one changes GRX provider, it keep the ASN. But if the private ASN is allocated from a LIR, it belongs to the LIR.

A Public ASN is necessary for the GRX providers in order to exchange routing information with their peers.

1.5.2.2 Connection of MNOs to the Internet Network

The GGSN or PDN Gateway is also connected to the internet network as shown ny Figure 1.10. The structure is the same as for the GRX with ASNs for the Internet Providers such as ASN 20771 (Caucasus Cable System).

1.5.2.3 How to Trace the Path to a Destination IP through the Various AS?

The "traceroute" utility which exists in the various UNIX versions, MAC OS, etc. sends ICMP (Internet Control Message Protocol) "Echo Request" to the destination IP. It sets a number of transit "hops" (also called PPEP) limit called TTL (Time-to-live) which is decremented at each hop. If it reaches 0, the hop returns an error message "ICMP Time exceeded" to the sender which contains the IP address of this hop.

The traceroute utility increases gradually the TTL until the final destination address is responding that the ICMP Time exceeded.

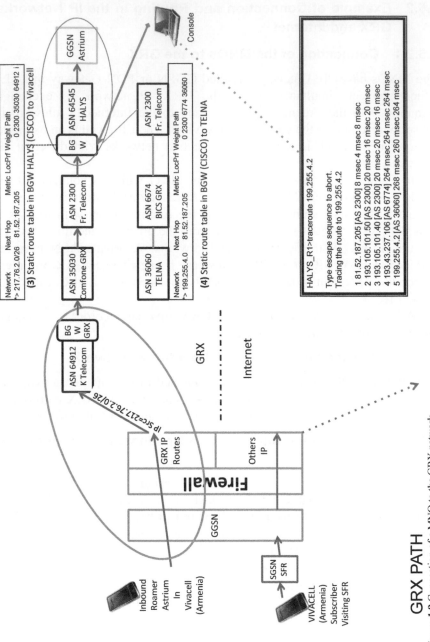

GRX PATH

Figure 1.9 Connection of a MNO to the GRX network.

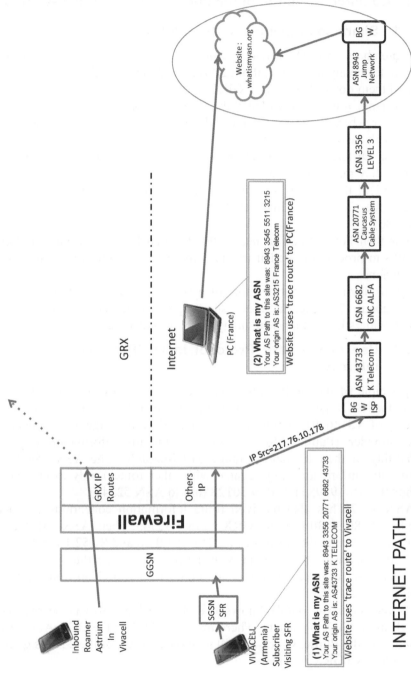

Figure 1.10 Connection of a MNO to the Internet network.

1.5.3   Common Structure of the Internet and GRX IP Networks

1.5.3.1   Sharable Equipments: the PPEPs

There are about 25 GRX providers *which are also carriers*. In the various GRX routes, many more AS (about 200) provide intermediate nodes where many of them do not have then a GRX provider role.

The main routers are called Public Peering Exchange Points (PPEP). A provider which is both Internet and GRX provider may use the same PPEP, with two different IP addresses for the accesses with the two networks. Different routing tables isolate logically the two networks although they may have many common equipments (large PPEPs). We see in Figure 1.11 that there is a common PPEP (name = LINX Juniper) shared between the Google AS(Internet ASN=15169)   and the KPN AS(GRX ASN=286).

It is then not difficult to use a PPEP which has routes with the two networks to process the transit traffic coming from mobiles or from fixed lines stations.

By displaying the routing table of a BGW on the GRX network, one can see about 170 different ASNs, showing that the maximum number of GRX providers is less than 170, many GRX providers having several AS.

1.5.3.2   Use of the BGP Protocol in an IP Network, GRX or Internet

The principle of the BGP route advertisement and how it « percolates » trough the GRX network is shown in Figure 1.9. The Private Border Gateway of ASN 64912 sends an "advertize" BGP message to his adjacent GRX provider (Public ASN 35030). It contains information on the reachability of their IP address range 217.76.2.0/26. The ASN 35030 is relaying this information (aggregated with other routes) only to his GRX neighbors ASN 19395 and ASN 30128 (not to ASN 52345).

ASN 30128 has also received route information concerning another route 81.52.180.92/30 from his GRX partner 52345. Then, ASN 30128 can advertise the availability of routes 83.217.254.0/26 and 81.52.180.92/30 to the Border Gateway of his client with the private ASN 64681. This ASN 64681 had also received the info concerning 81.52.180.92/30 from another supplier ASN 52345.

The Border Gateway 64681 knows that 81.52.180.92/30 can be reached through his 2 GRX providers. There are "weights" in the BGP Advertise message and the Border Gateway will take the shortest of the 2 paths going through his two GRX providers.

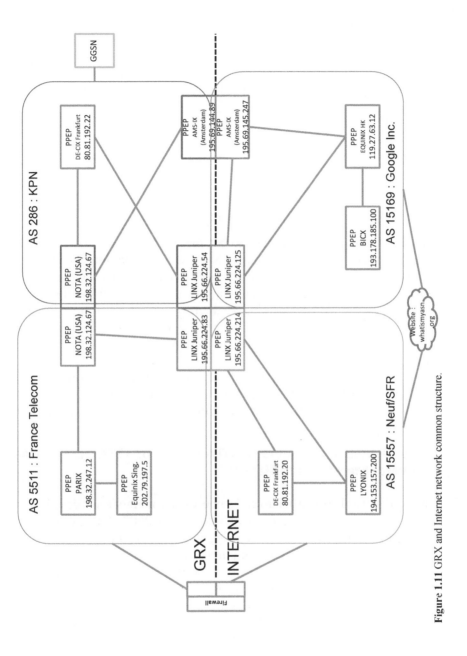

Figure 1.11 GRX and Internet network common structure.

In a Border Gateway, there is a "route table" giving for each IP range (a GPRS partner), the ASN of the neighbor GRX (in general only one in most cases), which is built by the reception of BGP messages/ The number is not huge as there are about 1400 mobile networks with in general at most 3 ranges for their internal equipment IP addresses.

The BGP is described in [1.4] and [1.5] and includes five types of messages:

1. OPEN
2. UPDATE
3. NOTIFICATION
4. KEEPALIVE
5. ROUTE REFRESH

1.5.3.3   GRX is an "Open Network"

All the IP routes are broadcasted to all the AS connected to the GRX whether private or public. Much like in SS7, without Roaming Hubs which may have detailed barrings depending on origin and destination networks, eventually origin, destination and auxiliary IMSI used (multi-IMSI service). The control of the GPRS roaming is then essentially performed by the firewalls in Figure 1.9. When a new range of equipments IP addresses of a partner is added (through the IR1), the VPLMN must add them in the firewall.

This requires work, and some MNOs prefer to have a firewall fully "open" for all the forward IPs They control the GPRS roaming only with the E212->E214 tables in the SGSN for UPDATE LOCATION GPRS.

1.6 The Device Management (OTA-GPRS or OTA-TDSA) Has Several Standards to Handle

When we discuss LBO in Chapter 3, one of the methods proposed is to change the APN of the visitors to EUinternet. In 2009 [0.8], it could be said that most of the handsets were using the OMA standard and only the OMA generic profile was covering most of the handsets. With more smartphones, this is no longer true. It is also true that there is not a single standard for Android handsets: Samsung's use the regular OMA while HTC has its own standard. And an OMA profile sent to an HTC phone is just ignored. The corresponding "load and test" OTA method of [0.8] would have five generic standards (2013) to try instead of two (2008):

- Nokia-Ericsson (rather obsolete)
- OMA

- HTC
- iPhone
- Blackberry

In the example below, we have three profiles models for internet access in the same network but using different standards.

It must be noted (2013) that some new smartphone software no longer have the standard GSM capability of working with a SIM Tool Kit in the SIM card. As certain updates are automatic this creates unwanted service impossibilities.

1.6.1 Reminder: Internet Profile with the Nokia-Ericsson Provisioning Standard

This method will use two SMS sent to port 49999

```
?xml version="1.0"?
!DOCTYPE wap-provisioningdoc PUBLIC "-//WAPFORUM//DTD PROV
1.0//EN""http://www.wapforum.org/DTD/prov.dtd"
wap-provisioningdoc
characteristic value=""
parm name="COUNTRY" name="MAX-NUM-RETRY"
/parm
parm name="NETWORK" name="NAPID192.168.10.100"
/parm
parm name="INTERNET" unknown=""
/parm
parm name="STARTPAGE" name="NAPIDorange"
/parm
parm name="TO-NAPID" unknown=""
/parm
parm name="PORTNBR" name="NAPIDorange"
/parm
parm name="SERVICE" name="NAPIDorange"
/parm
parm name="PROVURL" name="NAPIDorange"
/parm
parm name="PXAUTH-TYPE" name="NAPIDorange"
/parm
/characteristic
characteristic name="NAME" name="NAPIDhttp://www.orange.fr"
/characteristic
characteristic name="NAP-ADDRESS"
parm name="PROXY-ID" name="NAPIDOrange World"
/parm
/characteristic
characteristic unknown=""
parm name="PROXY-ID" name="NAPIDOrange World"
```

/parm
/characteristic
/wap-provisioningdoc

1.6.2   Reminder: Internet Profile with the OMA Provisioning Standard

This uses four SMS sent to port 2948 as the XML has the compact WBXML coding. Several Android smarphones use this standard and an OMA generic profile is applicable as well to most Nokia, Motorola, etc.

```
?xml version="1.0"?
 !DOCTYPE wap-provisioningdoc PUBLIC "-//WAPFORUM//DTD PROV 1.0//EN"
"http://www.wapforum.org/DTD/prov.dtd"
wap-provisioningdoc version="1.0"
characteristic type="BOOTSTRAP"
parm name="NAME" value="Orange World"
/parm
/characteristic
characteristic type="PXLOGICAL"
parm name="PROXY-ID" value="Orange World"
/parm
parm name="NAME" value="Orange World"
/parm
characteristic type="PXPHYSICAL"
parm name="PHYSICAL-PROXY-ID" value="WAPGateway"
/parm
parm name="PXADDR" value="192.168.10.100"
/parm
parm name="PXADDRTYPE" value="IPV4"
/parm
parm name="TO-NAPID" value="NAP_Orange World"
/parm
characteristic type="PORT"
parm name="PORTNBR" value="8080"
/parm
/characteristic
/characteristic
/characteristic
characteristic type="NAPDEF"
parm name="NAPID" value="NAP_Orange World"
/parm
parm name="BEARER" value="GSM-GPRS"
/parm
parm name="NAME" value="Orange World"
/parm
parm name="NAP-ADDRESS" value="orange"
/parm
```

```
parm name="NAP-ADDRTYPE" value="APN"
/parm
characteristic type="NAPAUTHINFO"
parm name="AUTHTYPE" value="PAP"
/parm
parm name="AUTHNAME" value="orange"
/parm
parm name="AUTHSECRET" value="orange"
/parm
/characteristic
/characteristic
characteristic type="APPLICATION"
parm name="APPID" value="w2"
/parm
parm name="TO-PROXY" value="Orange World"
/parm
parm name="NAME" value="Orange"
/parm
characteristic type="RESOURCE"
parm name="NAME" value="Orange"
/parm
parm name="URI" value="http://www.orange.fr"
/parm
parm name="STARTPAGE" value="http://www.orange.fr"
/parm
/characteristic
/characteristic
 /wap-provisioningdoc
```

1.6.3 Internet Profile for the HTC XML Standard (Same for Their Window and Android OS-Based Handsets)

They do not use the compact 1 octet tag of the OMA standard [1.11], but instead full character tags such as GPRSInfoAccessPointName. As a result the internet profile below uses 10 SMS sent to the port 2948 of the OMA standard. There are unuseful settings such as the obsolete Circuit Mode data access setting CM_PPEntries, only the GPRS setting CM_GPRSEntries is useful.

```
<?xml version="1.0"?>
<!DOCTYPE wap-provisioningdoc PUBLIC "-//WAPFORUM//DTD PROV 1.0//EN"
"http://www.wapforum.org/DTD/prov.dtd">
<wap-provisioningdoc>
<characteristic type="CM_Networks">
<characteristic type="Connection 1">
<parm name="DestId" value="{0AA202B8-13D2-13D2-F0CD-000028AD5DF3}"/>
```

```
</characteristic>
</characteristic>
<characteristic type="CM_PPPEntries">
<characteristic type="Orange World">
<parm name="DestId" value="{0AA202B8-13D2-13D2-F0CD-000028AD5DF3}"/>
<parm name="Enabled" value="1"/>
<parm name="CountryCode" value=""/>
<parm name="AreaCode" value=""/>
<parm name="Phone" value="+33674501200"/>
<parm name="UserName" value="orange"/>
<parm name="Password" value="orange"/>
<parm name="Domain" value=""/>
</characteristic>
</characteristic>
<characteristic type="CM_GPRSEntries">
<characteristic type="Orange World">
<parm name="DestId" value="{0AA202B8-13D2-13D2-F0CD-000028AD5DF3}"/>
<parm name="Enabled" value="1"/>
<parm name="UserName" value="orange"/>
<parm name="Password" value="orange"/>
<parm name="Domain" value=""/>
<parm name="SpecificIpAddr" value="0"/>
<parm name="IpAddr" value=""/>
<parm name="SpecificNameServers" value="0"/>
<parm name="DnsAddr" value=""/>
<parm name="AltDnsAddr" value=""/>
<parm name="WinsAddr" value=""/>
<parm name="AltWinsAddr" value=""/>
<characteristic type="DevSpecificCellular">
<parm name="BearerInfoValid" value="1"/>
<parm name="GPRSInfoValid" value="1"/>
<parm name="GPRSInfoProtocolType" value="2"/>
<parm name="GPRSInfoL2ProtocolType" value="PPP"/>
<parm name="GPRSInfoAccessPointName" value="orange"/>
<parm name="GPRSInfoAddress" value=""/>
<parm name="GPRSInfoDataCompression" value="1"/>
<parm name="GPRSInfoHeaderCompression" value="1"/>
<parm name="GPRSInfoParameters" value=""/>
</characteristic>
</characteristic>
</characteristic>
</wap-provisioningdoc>
```

1.6.4   Blind OTA Provisioning Strategy

When one attempts to use a standard unsupported by a given model, the settings are ignored and the arrival of a configuration SMS is not even shown. There is then no problem to load all the three MMS profile types, only one will be taken by the handset. The "loop feedback strategy" [0.8] which consists to send an MMS is then applicable, and even the handset model and IMEI can be seen in the MMSC traces as shown below.

1.7 Mobile Number Portability Implementation Solutions

Mobile Number Portability is a classical subject as it was implemented first in Hong Kong (1999) to ease the competition between their six MNOs as it allows a customer to change operators while keeping his number. It allows the porting of both voice and *non-call related signaling services* such as SMS. The solution in Hong Kong used an implementation at the level of each MNO, that is, the carrier which the number range belongs to receives the call or the SRI_FOR_SM request. If the number is ported out, the donor network is able to relay the call or the SRI_FOR_SM to the new network by prefixing the MSISDN with the new network prefix (*Routing Number*).

This service is also widely used in Europe where the MNP is effective since 2005 and in North America since 2002.

1.7.1 Strategies for the MNP Implementation in a Country and Variations

In most countries, the implementation of the MNP is a decision of the telecom regulatory body as it is not in the interest of the existing MNOs to facilitate the "churn" of customers which is costly. The implementation may be more (direct routing) or less (indirect routing) complicated as explained with the help of Figure 1.12 for the case of calls received in the ported case. Figure 1.12 shows three MNOs in the same "MNP domain", calls from/to international and fixed lines are outside this MNP domain.

Direct routing (a) means that, when a customer from MNO **B** calls a subscriber of MNO **A** which was originally from MNO **C**, the call goes directly to MNO **A**. There is no *tromboning* to the Number Range Holder (NRH) MNO C.

There are three technical strategies for the implementation at the various MNOs:

- Use of a central MNP database with synchronized copies at each MNO. This is a publicized strategy because it provides the direct routing. *But it is not at all mandatory* and not very cost-effective as the porting rate is low.

- A national MNP Gateway for all inter-operator and international calls and SMS. This is the implementation in North America.

- No centralized database. Each MNO only needs (many HLRs have the feature) to know for ported-out numbers, the "Routing Number" (RN) of the first network to which their customer was originally ported. They are not concerned by what happens after until (it happened rapidly in Hong Kong) he comes back. In this case, they just take out from their HLR the RN, and all ported-out memory is erased. Of course, only indirect routing is provided (the Number Range Holder operator is always involved) but it is very simple and cost effective. If the porting rate is 5% /year, the case of an interest for the improved direct routing is 5% of the calls; with a low proportion of about 0.25% for 2 indirections of calls or SMS. This is called the "Onward Routing method" in Table 1.5.

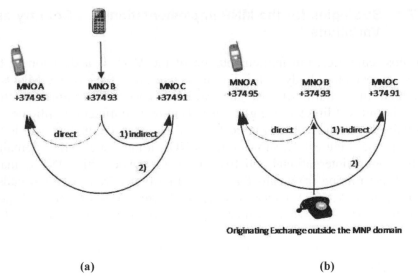

(a)   (b)

Figure 1.12 (a) Direct and indirect routing for outgoing calls (b) for incoming calls from outside the MNP domain.

1.7.1.1 Role of a MNP "Clearing House"

It *may* be decided to create a central MNP database (MNPDB) which is administered by an independent MNP Clearing House (MCH), which could be a joined venture of the MNOs to make it available to all the MNOs. *Simpler MNP implementations do not require this.* The customers who want to change their operators contact the new operator that they have selected, which will through the MCH channel, make the change to the central MNP database. This database or a daily synchronized copy is used by the MNOs and after a short delay, the customer will be using his new operator. The DB structure is not standardized and country dependent, an example of a customer record is given in Table 1.4.

To secure the porting requests which are submitted at one of the recipient network's sales counter, the Customer Identifier may be requested by the regulatory body. A central IVR or USSD service accessing the MNPDB may be supplied by the MNP Clearing House (or each MNO). The Identifier is delivered based on the MSISDN which is still operational at this stage in the donor network.

1.7.1.2 Number Portability Implementation Solutions for the Concerned MNOs

Each MNO may have their own implementation to use the database, two architectures are defined in [1.7]:

- MNP-SRF (using a MNP Signaling Relay Function), this can have also various implementations. This is the most popular solution as it handles both call and non-call related services such as SMS,
- ANY_TIME_INTERROGATION, etc. using the MSISDN to identify a customer.
- IN-based (using a MNP script in the IN system of the MNOs), *not used for mobiles because it does not provide SMS with standard MSCs,* (an SMS from an external SMSC interrogates directly the HLR and there is no trigger for a Camel InitialDPsms), but used extensively for fixed line call services which do not include SMS.

The cost of the central database and of the third party structure is relatively expensive. For small countries which have MNP projects, it is possible to have the MNP service working without any central database administration. An MNP-SRF solution implemented since several years which create the database dynamically will be explained.

Table 1.4 Example of MNP Database Record

Name of the parameter	Content
MSISDN (Mandatory)	Mobile Number which will be kept by the customer
RN (Mandatory)	Routing Number (prefix identifying the current MNO of this customer, e.g. could be the MGT of the MNO)
Customer Identifier (Optional)	Country-specific (assigned to a customer when he gets a MSISDN, it is used to secure his porting request). He can obtain it through USSD or an IVR from his current operator.
Donor network (Optional)	Network where the number was subscribed to before porting.
Timestamp (Optional)	Date of porting

There are three main routing methods to use MNP ([1.8] gives other methods rarely used):

- All Call Query: the MNP-SRF is systematically invoked for all calls and SMS, this is the most usual case. If the All Call Query is used, the number ranges of the operator set in the GMSC for HLR enquiry are simply replaced by the range of all mobile numbers of the country
- Query on release: the portability entity is queried only when a subscriber MSISDN is not found in a MNO HLR. If the rate of "porting" is low, the criticality of the portability entity is reduced.
- Onward Routing: that method can be used with only a private database containing only ported-out numbers of the MNO. In this case, the call is relayed to the next operator known for a MSISDN (and so on)

The first implementations did not consider important to avoid the simple initial relay for outgoing calls or SMS-MT from a mobile to a « ported-out » number, first to the original carrier, then to the subscribed carrier. But direct routing provides a more efficient call routing.

Table 1.5 Number Portability implementation solutions

		Standard solutions	
Routing type	Main routing methods	MNP-SRF (Signalling Relay Function) Call and non-call related	IN (Intelligent Network) Call related only
Direct routing	All Call Query	Direct routing	Originating Query on Digit Analysis (OQoD)
Indirect routing	Query on Release	N.A.	Query on HLR Release
	Onward Routing	Indirect Routing with reference to subscription network	Terminating Query on Digit Analysis (OQoD)

1.7.2 Description of the MNP-SRF Solution with Synchronized MNPDB Copies

1.7.2.1 Call Flow Principle with a MNP-SRF Solution

This MNP-SRF solution is described in [1.7] but Figure 1.13 unifies the ported and not ported cases.

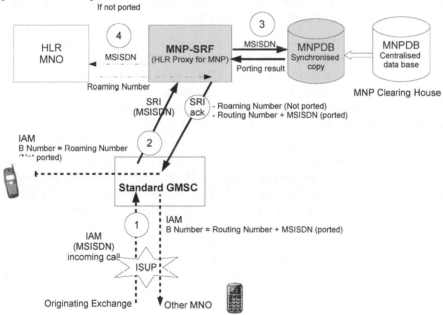

Figure 1.13 Incoming call with the MNP-SRF solution.

1.7.2.2 SMS Case

The call-flow is similar to the one above except that, for ported numbers, the SRI-for-SM (and not the IAM) is relayed by the MNP-SRF to the proper HLR (using the Routing Number as a prefix of the MSISDN in the SCCP Called Party).

1.7.2.3 Implementation in the HLR

A popular implementation is by the HLR supplier. The MNP-SRF including the MNPDB is integrated in the HLR. It is a rather cost effective solution and most vendors have it. For older software versions, at least the "Onward routing" indirect method is available.

1.7.2.4 Implementation in the GMSCs

Another cost effective solution is to have the MNP-SRF and the MNPDB included in the GMSC as the GMSCs are the switch point for SCCP and ISUP between an MNO and his roaming partners or local partners. Several GMSC vendors propose an implementation of MNP in their GMSCs which are able to synchronize with the central database.

1.7.2.5 Separate MNP-SRF Equipment, Case with an Automatic Database Creation

There is an alternative solution which works *without a static central database or* copies and with the standard MSC setup (no need for "features"), provided that the various MNOs have SS7 links between them. In Figure 1.14, a call is received (1) As in the case with a copy of the MNDB, the GMSC was configured to interrogate the MNP-SRF for all incoming national B-numbers (both the MNO and *all his competitor's ranges*). So the GMSC sends a SEND_ROUTING_INFO to the MNP-SRF with the B-Number (2), which interrogates his "cache database"(3) which is assumed to be empty initially. If no entry is found, the proxy MNP-SRF interrogates first the real HLR of the MNO (4) (Point Code addressing is used to avoid a loop with the path HLR – MNP-SRF through the GMSC), then if not found, *sequential SEND_ROUTING_INFO are sent to the HLRs of his competitors* (5, 6, etc.) until the B-Number is found. Then the MNP-SRF responds SEND_ROUTING_INFO ack (SRI ack) as in the previous case 1.7.2.1 with either the Roaming Number of the called party if it belongs to the MNO or Routing Number+MSISDN if it is ported to another MNO.

The MNP record of Table 1.4 for the B-Number *is added in the cache automatically* by the MNP-SRF, so no search will be necessary next time,

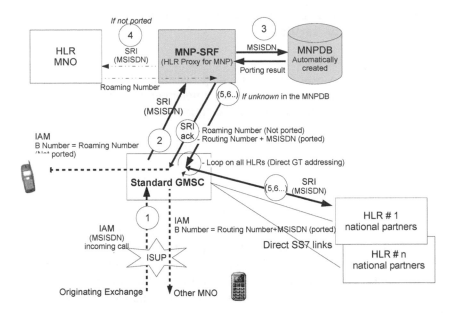

Figure 1.14 Incoming call with the MNP-SRF solution without a centralized database.

only a check that the number is still in the cached MNPDB.

An outgoing call (IAM with Routing Number+MSISDN) is sent by the GMSC if the B-Number is ported-out.

For the SMS, it works the same as explained previously for the case with a synchronized MNPDB, SEND_ROUTING_INFO_FOR_SM are used instead of SEND_ROUTING_INFO.

1.7.2.6 Provisioning of the Mobile Number Portability in the MNOs with the Automatic MNPDB Solution

There is no need of a central database, the MNO which the number is ported-out from will simply suppress the number from its HLR (which will then return « unknown number » when interrogated; the MNO which the number is ported-in to will add it in its HLR.

1.7.2.7 Case When There are No Direct SS7 Links between the Operators

The solution is almost the same but the central database cost cannot be avoided. The cost saving comes from the HLR proxy if it is cheaper than the MNP software in the GMSC.

1.7.3   Description of the IN Solution for the Fixed Lines Portability Service

A frequent case without a change of operator for fixed line services is when someone moves and wants to keep his number in a different region. Some countries have more expensive inter-operators call cost when an outgoing "prefixed call" is relayed. A "query" may then be added for all outgoing calls to allow direct sending to the subscribed network without going through the original network if it is different.

A query is performed by the MSCs or the fixed switches, before any outgoing calls, to a SCP (INAP CS1) with an extension to manage portability (call it "IN Proxy" in Figure 1.15). In the MSCs or the fixed networks switches with INAP CS1, "network Initiated triggers", (N-CSI in INAP CS1) are configured. For each outgoing call, an INITIAL DP message is triggered. The B-Number is processed by the IN proxy (query of the MNPDB) and a prefixed number is created. The IN Proxy will force the direct call to the subscribed network if the B-Number is ported out with a CONNECT (RN+Called Party Number). If not, the IN Proxy just returns a CONTINUE on the INAP dialogue.

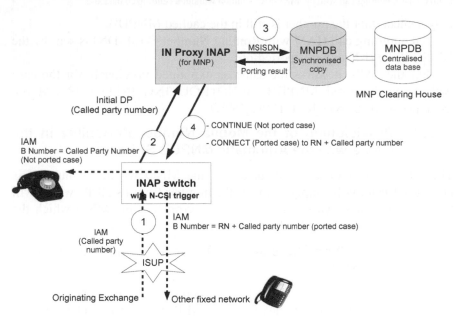

Figure 1.15 IN solution for the fixed lines portability service.

This solution for fixed line portability is easily included in the fixed line switches which all have INAP since more than 10 years. Of course, it could be used also for the voice service in mobile networks but is not used because it does not provide the SMS service.

1.7.4 Details of the MAP-ISUP Mapping for the Implementation of the MNP in the MNOs

On the MSC side, a proper implementation of MAP v1, v2 or v3 is the only feature is required to implement one of the following addressing modes. The Optimal Routing feature in the GMSC is not required. The various MSCs (including GMSCs) of a MNO, when there is no MPNP in the country, when there is a call, make a SEND_ROUTING_INFO Req to the HLR *only when the destination number is in their range.* Implementing MNP is then simply extending the table to make the SRI request to the MNP-SRF for any national mobile number.

1.7.4.1 Concatenated Addressing

In MAP SRIack messages from NPLR, MAP versions 1 and 2 only support concatenated addressing for MNP. That means that MSRN parameter in the SRIack will contain the MSISDN prefixed with the Routing Number (RN). Then, the IAM *Called Party Number* parameter includes both RN and MSISDN concatenated.

1.7.4.2 Separated Addressing

In MAP v3, the MNP indicator and the MSISDN parameters in SRIack allow to have the Routing Number alone in the MSRN parameter if the number is known to be ported. In this case, the IAM *Called Party Number (B Number)* parameter contains the Routing Number and the *Called Directory Number* (CDN) parameter of MAP contains the MSISDN. The SRIack contains also a MNP indicator to tell the GMSC that it must route the call to the other network with the B Number=RN + CDN.

Other methods may be used (see [1.10]), but in all cases, the method to transport the routing number in the IAM depends on the interfaces agreed by the operators in the portability domain.

1.7.5   Implementation of Different Rates for Calls to Ported-out Mobile Numbers

Some MNOs depending on the country regulation may wish to implement different rates for calls from their network to other networks including the ported-out numbers.

For prepaid subscribers, the gsmSCF controlling the call may send prior to the call an Any Time Interrogation (ATI) message to the MNP-SRF in order to know the portability status of the destination number. If the destination number is ported, the MNP-SRF will relay the ATI to the correct MNO (using the Routing Number as a prefix of the SCCP Called Party). The gsmSCF will then receive the response to the ATI and may apply a different rate according to the destination IMSI or Routing Number

To explain the details of how it is possible for the IN to charge a different rate for a call to a mobile which does not belong to the same network as in Figure 1.12 (a), here is the trace of the ANY_TIME_INTERROGATION request sent by the IN of MNO B to the NPLR (which has the MNP-SRF function and is an HLR proxy). The information requested concerns only the MNP status.

```
- - - - Super Detailed SS7 Analyser (C)HALYS (Trace Level 8) - - - - - - -
    PA_Len = 30
    MAP-ANYTIME-INTERROGATION-REQ(29)
     MAPPN_timeout(45)
      L = 002
      Data: timeout value =  10 sec
     MAPPN_invoke_id(14)
      L = 001
      Data: 1
     MAPPN_req_info(53)
      L = 001
      Data: 03
          location information  not requested
          subscriber state      not requested
          current location      not requested
          IMEI                  not requested
          MS classmark          not requested
          MNP requested info  requested /* only info requested !! */
    MAPPN_gsmscf_addr(51)
     L = 007
     Data: Ext = No extension
          Ton = International
          Npi = ISDN
          Address = 37493297200    /* GT of the IN (SCF) */
    MAPPN_msisdn(15)
     L = 007
```

```
Data: Ext = No extension
    Ton = International
    Npi = ISDN
    Address = 37493123456  /* MNP status requested for this number */
```

The MNP-SRF returns that the number has been ported out to another network:

```
- - - - Super Detailed SS7 Analyser (C)HALYS (Trace Level 8) - - - - - - -
PA_Len = 39
MAP-ANYTIME-INTERROGATION-CNF(152)
 MAPPN_invoke_id(14)
  L = 001
  Data: 1
 MAPPN_Routing_Number(550)
  L = 001
  Data: Ext = No extension
      Ton = International
      Npi = ISDN
      Address = 37495 /* Routing Number of current network */
 MAPPN_mnp_msisdn(567)
  L = 007
  Data: Ext = No extension
      Ton = International
      Npi = ISDN
      Address = 37493123456  /* MNP status requested for this number */
 MAPPN_mnp_number_portability_status(568)
  L = 001
  Data: (1): own number ported out
```

The IN may charge a rate higher than for a call to a mobile belonging to the concerned MNO. For the postpaid subscribers, the simplest is usually to declare them also with a Camel profile so that their outgoing calls trigger the SCF.

References and Further Reading

(The 3gpp standards can be downloaded free at www.etsi.org)

[1.1] GSMA, IR.80 V1.2, "Technical Architectures for Open Connectivity Roaming Hub Models".
[1.2] 3gpp TS 23.00, V 11.5.0 (2013-04), "Universal Mobile Telecommunications System" (UMTS); Numbering, addressing and identification, (Release 11)", a most useful document which gives all the codings used in the mobile protocols, MAP, CAMEL, Diameter, BSSAP.
[1.3] RFC 4271, "A Border Gateway Protocol 4 (BGP-4)", Jan. 2006.

[1.4] RFC 2918, "Route [Refresh for BGP-4, September 2000.

[1.5] 3gpp TS 29.002, V 11.5.0(2013-11), "Mobile Application Part (MAP) specification, Release 11".

[1.6] GSMA, IR.75 V2.0, "Open Connectivity SMS Hubbing Architecture".

[1.7] 3gpp TS 23.066, V 11.0.0(2012-6), " Digital cellular telecommunications system (Phase 2+); Universal Mobile Telecommunications System (UMTS); Support of Mobile Number Portability (MNP); Technical realization; Stage 2, Release 11".

[1.8] GSMA, MNP White Paper, March 2009.

[1.9] 3gpp TS 23.079, V 11.0.0 (2012-10), "Universal Mobile Telecommunications System (UMTS); Support of Optimal Routeing (SOR); Technical realization, Release 11".

[1.10] ITU-T Q769.1, ISN user part enhancements for the support of number portabilility.

[1.11] Open Mobile Alliance, « Client ProvisioningV1.1 » , July 2009

2

Principle of Virtual Data Roaming
Architectures

NSK   *"Все что наше наше. Все что ваше договорный." – "All which is ours is ours, all which is yours is negotiable."*

JFK   *"The USA has a huge power, we can completely destroy you 30 times!"*

NSK   *"The Soviet Union does not waste his people's money, we can completely destroy you only once, it is optimal."*

       "Мы вам покажем кузькину мать!" – "We shall show you Kuska's mother!"

       – J.F.Kennedy and N.S. Krutschev private peace talks,1961.

2.1 GTP Hubs GSM and 4G LTE

The GTP protocol is common between GSM and LTE, versions 0 and 1, version 2 for 4G as well as for recent 3G. The GGSN is called PDN Gateway in LTE and the MME of LTE performs the SGSN and RNC combined roles.

2.1.1 Case of a single GTP Hub

The case of Figure 2.1 is that of a Bi-IMSI (or multi-IMSI) virtual roamer for GPRS data services. He is visiting the SGSN or MME (4G LTE) on the right using one of the auxiliary IMSI because his nominal IMSI does not have roaming with this VPLMN. So the Bi-IMSI Applet in the SIM card has automatically selected an auxiliary IMSI which is able to register successfully.

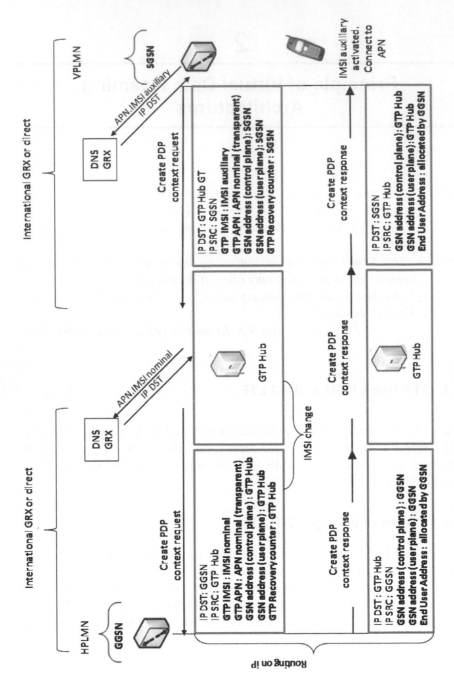

Figure 2.1 Principles of GTP hub for Bi-IMSI with one roaming hub.

There are four commercial parties involved in providing the service:

- the "client network" which has a GGSN (or PDN Gateway 4G) on the left (.e.g. Mexico network),
- the sponsor network which has provided a range of auxiliary IMSIs to the GTP Hub provider which allows to roam (e.g. sponsor is a European network),
- the VPLMN on the right (e.g. A French network),
- the GTP Hub service provider.

2.1.1.1   Commercial Roaming Setup for the Service

The client (Mexico) is a client of the GTP Hub which is providing him the Bi-IMSI service for data. The GTP Hub has its own GRX IP address (assume there is only one) that can use the GRX network and has requested the sponsor to include it on his IR21 (as one of his own GGSNs).

The client has also provided the full list of his APNs. His supplier, the Roaming Hub EXIT is asking the sponsor regularly to include all of them also on his IR21. This is because of the control performed by some VPLMN which block unlisted APNs, while others do not allow wild carting ("*" is set as APN in the GPRS profile sent by the HLR or HSS to the VPLMN).

2.1.1.2   Routing Information Elements in the GTP Protocol and Mechanism

In the VPLMN SGSN, the address resolution with the DNS uses the sponsor's IMSI (auxiliary IMSI) with the following argument:

mms.itelecel.com.mnc840.mnc310.gprs (e.g. Telna is the sponsor)

The DNS will give the IP address of the GTP Hub *if its IP address and the APN have been entered* in the DNS used by the visited SGSN. The "Create PDP context" message will be sent by the SGSN to the GTP Hub.

The only thing which allows the GTP Hub to identify to which client network to address the initial GTP message "Create PDP Context" is not the Destination IP address (always its own), but the IMSI contained in the IMSI Information Element (IE) of the GTP message. If the auxiliary IMSI received does not match one of the nominal IMSI, the GTP message would be relayed to another GTP Hub (see Section 2.2).

Otherwise the GTP Hub does:

auxiliary IMSI (e.g. 425019876543210 -> nominal IMSI (e.g.334209988776655) and interrogates his DNS by using also the APN received with the argument:

mms.itelecel.com.mnc020.mnc334.gprs

(e.g. Telcel Radio Mobil Mexico MMS service).

then sends the Create PDP context to the IP address of the GGSN of the HPLMN.

As you can see, IMSI and APN are the routing parameters used in addition in the GTP Hub.

The GTP Hub is the home GGSN for the SGSN. The Create PDP context received from the SGSN is resent to the home GGSN *with a change in the GSN address (control plane) and the GSN address (user plane)* which are the IP addresses to which the real home GGSN will respond. The GSN address (user plane) should be changed so that the *GTP-U data (for example HTTP requests) go through the GTP Hub*. It is necessary in general because, although the SGSN and the GGSN are both on the GRX network, the visited network will not have opened the IP address of the GGSN (no data roaming agreement assumed) in his "Border Gateway" firewall.

2.1.2   Case of a Chain of Peering GTP Hubs

When a GTP Hub wants to open a new VPLMN, it may take several months before such VPLMN implements its IP address found in the sponsor's IR21. If it can use an IP address already opened from the VPLMNs' side, it will be much quicker. This is explained in Figure 2.2. In particular, it could use one of the sponsor's spare IP addresses. A sponsor involved in virtual data roaming should have a range of spare IP addresses in its IR21 that it can allocate to new GTP Hub service suppliers.

In this case, the GTP Hub ENTRY point is at the sponsor and is a simple IP router which routes the received Create PDP context to the Roaming Hub EXIT (that of the service provider): any message that is received to the IP address allocated to the partner GTP Hub is routed to the GTP Hub EXIT address (easy to open, this is a direct commercial relation).

More generally, it could also be an IP address of a peering GTP Hub supplier which has a full Roaming Hub opened with the VPLMN and that the Roaming Hub EXIT is partnering with.

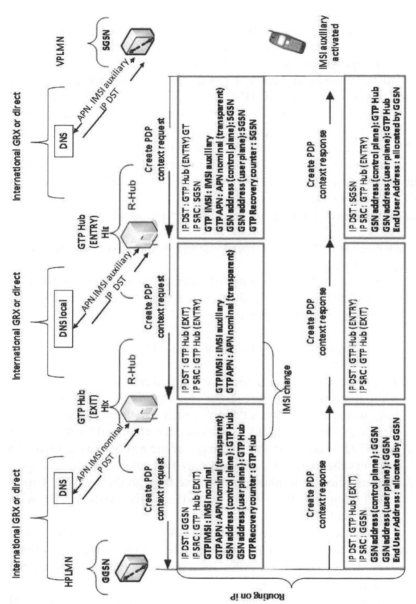

Figure 2.2 Principles of GTP hub for Bi-IMSI with a chain of 2 peering data roaming hubs.

2.2 Optimization with "local break-out(LBO)" with a Network of Cooperating Multiple GTP Hub

As explained in the introduction, a GTP Hub can provide LBO optimization with "local dynamic break-out" and "dynamic local break-out", without requirements to change the APN.

As will be detailed in Figure 3.1, the classical internet access for a visitor follows a path SGSN->GTP Hub -> GGSN home of the visitor. If the website accessed over internet is in the visited country, the latency may be quite long. It can also be expensive (roaming data charges average > 10 USD / Mb outside a privileged zone as Europe in 2013).

To improve the situation, the break-out consists in allowing a direct internet access through the GTP Hub without "tromboning" the data through the HPLMN GGSN.

The following setup modes in the GTP Hub are possible:

- a white list of IMSIs, all these visitors would have the "break-out" applied to them
- a white list of APNs, such as "EUinternet" as proposed by BEREC for July 2014, all the visitors using this APN would have the "break-out" applied to them

  (both cases are called "static local break-out")

- and, technically more difficult to implement, specific URLs in a white list would have "break-out" for any visitor, for example, the help web page of the VPLMN which can then advertise other "free websites" similar to what exists in certain WiFi hotspots. For Air GSM, it could be the web page of the airline thus allowing to suppress the expensive onboard duty free catalogue as passengers may use their smartphones to browse through the website. This is called "dynamic local break-out". It already exists in Air WiFi services in USA.

Chapter 3 explains the technical implementation of the "dynamic local break-out" which is more complex than the rather simple static local break-out based on IMSI ranges or APN lists, because the payload of the connection needs to be inspected.

To implement this, GGSN simple IP router function and the GGSN of the VPLMN are replaced by a GTP Hub in the VPLMN which has both functions. To avoid charging the visitors, SGSN tickets must be cancelled, so that they are not sent to the HPLMN. This is done by filtering with the

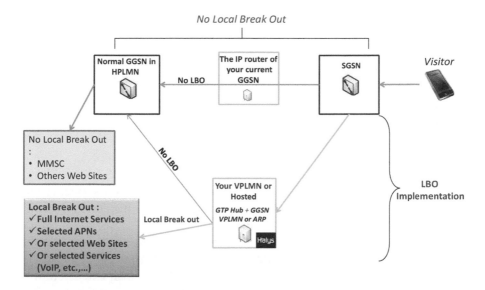

Figure 2.3 Principle of a dynamic Local Break-Out (LBO) service.

tickets created by the GTP Hub which include the IMSIs, APNs, Time stamps, etc. but also the website domain name such as "airfrance.fr", with a mention "break-out".

2.3 RADIUS Hubs

The issue here is "WiFi roaming", which allows a user with a WiFi subscription at home to use another WiFi network transparently in the same or in a foreign country

The term "WiFi roaming" may be misleading as the term is also used for the possibility to "handover" from an Access Point to another without losing the connection, which we will not cover.

2.4 SIP and RCS Hubs

RCS is a *pure IP service* designed to unify various mobile standard services:

- Standalone Messaging
- person to person instant messaging

- Group Chat
- File Transfer
- Content Sharing
- Social Presence Information
- IP Voice call (SIP)
- Video call
- Geo-location Exchange
- Network-based blacklist
- Capability Exchange based on Presence or SIP OPTIONS

To illustrate the service, when two users are talking (circuit service), one will turn on his camera and show the moving crowd to his correspondent: the packet services are available during a call.

The project is led since 2008 by the GSMA and is supported by several terminal vendors and software editors (a free client from Orange Labs is available). Basically, an RCS client can be considered as an extension of a SIP softphone and uses the SIP protocol mechanism for call setup and messages. Inside the networks, the most logical implementation is to use an IMS core.

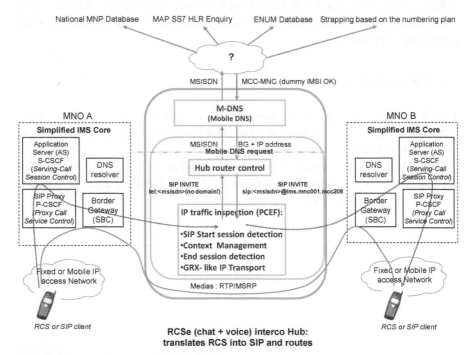

Figure 2.4 Message flow in the RCS Interco Hub.

Two subscribers of different networks may enter into a session. This is the RCS/joyn program using the RCSe specification of the softphone. It has been implemented notably between the three main Spanish operators. The main difference between the SIP and the RCS protocol is that RCS uses:

tel: <msisdn> (no domain)

to address the destination instead of

sip: <msisdn>@ims.mncXXX.mccYYY.gprs

Which means, especially in Europe that a "RCS interco Hub" must resolve the current network owning the subscriber, that is, determine the domain from the MSISDN. The use of the ENUM service has been specified for that by the GSMA, but such a service is not available in general (many operators do not want to provide their customer databases to others. Figure 2.4 describes a pragmatic solution using SS7 HLR queries and the search algorithm used in our SMS Hubs when there are several peers.

If the HLR is not available, an approximated resolution ("strapping") is done with the MSISDN number ranges. We use the same method as for our MMS Hub with a very similar problem: we must find the domain name of the IMS server instead of finding the MMSC.

The Message Session Relay Protocol (MSRP) is used to transfer related chat messages or for large file transfers during a session.

2.5 Diameter Hubs (LTE 4G Virtual Roaming Hubs)

2.5.1 Diameter Protocol Stack

Table 2.1 compares the protocols used in 2G-3G and 4G networks, while Table 2.2 compares more specifically the MAP and DIAMETER primitives. Table 2.4 compares the interfaces for Location Services explain in Chapter 13.

Table 2.1 *comparison of standards between 2G, 3G and 4G*

	2G, 3G	4G
Mobility Protocol	MAP TS 29.002	Diameter 29.272 interface S6a
Equipment control (Check IMEI)	MAP TS 29.002	Diameter 29.272 interface S13
Charging and Service control	CAMEL TS 29.078	Diameter Credit Control Use PCC (Policy Charging and Control)
Protocol Coding	ASN1	Diameter uses "Attribute Value Pairs(AVP)"

Table 2.2 *comparison of mobility and equipment checking messages between 2G, 3G and 4G*

Interface name 4G for mobility	2G, 3G and 4G LTE (S6d interface) TS 29.002	4G TS 29.272
S6a (MME) or S6d (SGSN) (Mobility Management)	UPDATE LOCATION GPRS SGSN -> HLR	UPDATE LOCATION MME or SGSN ->HSS
"	CANCEL LOCATION HLR->SGSN	CANCEL LOCATION HSS->MME
"	SEND AUTHENTICATION INFO SGSN->HLR	AUTHENTICATION INFORMATION MME or SGSN->HSS
"	INSERT SUBSCRIBER DATA HLR->SGSN	INSERT SUBSCRIBER DATA HSS->MME or SGSN
"	DELETE SUBSCRIBER DATA HLR->SGSN	DELETE SUBSCRIBER DATA HSS->MME or SGSN
"	PURGE MS SGSN->HLR	PURGE UE MME or SGSN ->HSS
"	RESET HLR->SGSN	RESET HSS->MME or SGSN
"	NOTIFY SGSN->HLR	NOTIFY MME or SGSN ->HSS
S13 or S13' (equipment control)	CHECK IMEI SGSN->EIR	CHECK IMEI MME or SGSN->EIR
S7a/S7d (Handover between Circuit Switched System an Packet Switched System)		UPDATE VCSG LOCATION MME or SGSN->CSS(Circuit Switched System)
"		CANCEL VCSG LOCATION CSS->MME or SGSN
"		INSERT VCSG SUBSCRIBER DATA CSS-> MME or SGSN
"		DELETE VCSG SUBSCRIBER DATA CSS -> MME or SGSN
"		VCSG RESET CSS-> MME or SGSN

We assumed that the reader is familiar with 3G virtual roaming using the Mobility Application Protocol MAP of GSM and CAMEL.

As LTE is only for data (no voice services or SMS), the mobility protocol is the subset of MAP corresponding to the SGSN originated or terminated signaling messages. It is far simpler than the MAP Mobility Protocol of 3G, 3G. The same message names have been used.

When addressing the HSS, the E214 address of 3G derived from the IMSI is replaced by the "EPC Home Network Realm". Chapter 6 will explain the necessary transformations in order to provide virtual roaming 4G for two networks that do not have a direct agreement.

S9 is the reference point interface between the PCC functions in the VPLMN and the HPLMN. The S9 protocol has the same role as CAMEL in 2G, 3G in very simplified terms.

The International Gateways are upgraded to route DIAMETER messages of the S6a and S9 protocol as they route SCCP messages in 2G, 3G (Diameter Signaling Routers). Some have the capability of translating S6a to MAP. The S7 and S13 messages are not normally conveyed in a roaming relation, only S6 and S9 are conveyed.

Table 2.3 Policy charging and control (PCC) messages in 4G

Interface name 4G for Policy Charging and Control(PCC)	2G, 3G	4G
S9 and Gx		Credit Control Request(CCR) V-PCRF -> H-PCRF and PCEF ->PCRF
"		Credit Control Answer(CCA) V-PCRF<-H-PCRF and PCEF<-PCRF
"		Re-Authorization Request(RAR) H-PCRF->V-PCRF
"		Re-Authorization Answer(RAA) V-PCRF->H-PCRF
"		Trigger-Establishment Request(TER) H-PCRF->V-PCRF
"		Trigger-Establishment Answer(TEA) H-PCRF<-V-PCRF

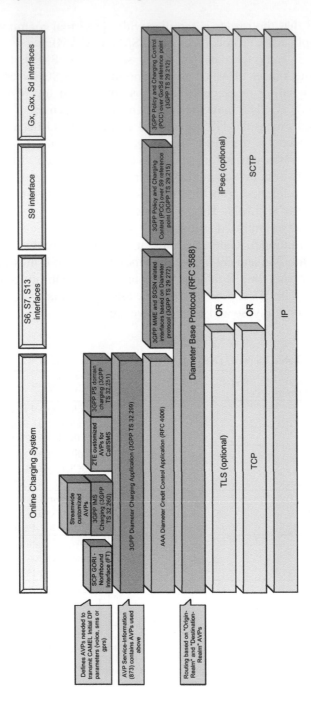

Figure 2.5 Diameter protocol based stacks.

Table 2.4 Location services messages in 4G

Interface name 4G for LCS	2G, 3G interface) TS 29.002 (MAP)	4G (DIAMETER)
SLh	SEND ROUTING INFO FOR LCS GMLC->HLR	SEND ROUTING INFO FOR LCS GMLC->HSS (TS 29.173)
Slg	PROVIDE SUBSCRIBER LOCATION GMLC -> MSC or SGSN	PROVIDE SUBSCRIBER LOCATION GMLC -> MME (TS 29.172)
SLs		TS 29.171

In the e-UTRAN LTE network, there are also the Location Services with specific DIAMETER- based primitives which are equivalent to their MAP or BSSAP-LE equivalents.

In Figure 2.5, we show some proprietary AVP implementation in light grey.

2.5.2   Different Types of Diameter Hubs

There are three types of roaming connections using Diameter. When an MNO gets a connection on the GRX network, he has a GRX public IP address which is the equivalent of an International Point Code (IPC) when using the SS7 (SCCP) network. This makes the creation of a roaming relation between two MNOs much simpler than with SS7: there is no need to build MTP routes as the GRX provides a route between any two subscribers of the GRX. They just have to open their Border Gateway (the equivalent for DIAMETER of a GMSC). Table 2.5 below compares the routing for roaming relations using SS7 or Diameter.

Table 2.5 Routing comparison between SS7 and Diameter roaming

	SS7 roaming	Diameter roaming
International Point Code (direct relation without the need of an International Gateway Provider)	International Point Code	GRX public IP address
Transparent relay of non routing related AVPs	International Gateway(SCCP)	Diameter Relay (defined in IR 88)
Inspect, may modify or add the AVPs (suitable for multi-IMSI roaming)	SS7 Virtual Roaming Hub	Diameter Virtual Roaming Hub Diameter Proxy (defined IR 88)

For example, in the case of a multi-IMSI roaming, the Diameter Proxy should perform

Auxiliary IMSI -> Nominal IMSI in the User-Name AVP (equivalent of IMSI). Here, a "Diameter Proxy" hub is necessary.

In the case of the "single IMSI virtual roaming", the GRX IP route and the commercial relation are not opened between a VPLMN and a HPLMN although they wish HPLMN's subscribers to be able to visit the VPLMN. The LOCATION UPDATE diameter is sent from the VPLMN to the Diameter Relay provider which is able to change the routing in order to reach the HPPLMN.

However, the mandatory AVP Visited-PLMN-Id (contains the MCC-MNC of the visited network) should probably be changed to one of the Roaming Hub Diameter to be accepted by the HPLMN.

In this case also, the "Diameter Proxy" mode is probably necessary. We use 'probably' because most Diameter Signaling Nodes (DSNs) are probably able to change some of the non-routing related AVPs.

The Diameter Relay mode probably does not correspond to any commercial practical case.

References and Further Reading

[2.1] 3gpp TS 29.272 v11.5.0, "LTE, Evolved Packet System (EPS), Mobility Management Entity (MME) and Serving GPRS Support Node (SGSN) related interfaces based on Diameter Protocol, Release 11", (description of the S6,S13 and S7 interfaces)
[2.2] GSMA, IR.88, V9.0, "LTE Roaming Guidelines".

3

GTP Hubs for GSM and LTE networks

3.1 Why GTP Hubs are Necessary in General

The GRX network is the backbone "Intranet" for data transmission between mobile operators. The communications between two operators are opened only when they sign an agreement. They use firewalls before their "Border Gateways" which have a list of Roaming Partner's GGSN IP addresses of their opened partners. There is a routing protocol called BGP that use ASN as a generic number defining the set of IP addresses of the mobile operator, similar to what is done on traditional IP networks.

A GTP Hub also allows carriers to open many new services described below: IP local break-out, APN correction using CAMEL SS7 alleviating the need to provision the mobiles, MMS anti-spam. The explanations and call flows will be given first for the simpler case of 3G networks using SGSN and GGSN, then we will present the System Architecture Evolution (SAE) of LTE and the adaptation of the solutions.

3.2 Single IMSI VPLMN's DNS Setup

This allows a single-IMSI data service for visitors who do not have a GPRS data agreement between the VPLMN and their HPLMN. The VPLMN IP addresses are not opened in the firewall of the HPLMN.

In the VPLMN (see Figure 2.2), the local DNS of the SGSNs is used to resolve the domain name of all the virtual visitors (e.g. mnc001. mcc208. gprs) with the IP address of the GTP Hub provider.

3.3 Multi IMSI Setup by the GTP Hub Operator

What type of setup has to be organized for his sponsors by the GTP Hub operator so that the data of the multi-IMSI users will flow through the GTP Hub while their auxiliary IMSI is active? If the GTPH does not have a GPRS roaming agreement with the VPLMN, messages outgoing from his own GRX IP address would be barred by the VPLMN's firewall *in general* (unless the VPLMN has an "open firewall GRX" which is rare), we will not assume the case.

Two methods are given in the following sections.

3.3.1   GTP Hub IP Included in the Sponsor's IR21

The GTP Hub operator agrees with his sponsors to:

(a) Include his own GTP Hub IP in their IR21 (see Section 1.3 of Chapter 1), which includes the list of all IP ranges (the firewalls will then be opened by the VPLMN);

(b) Reserve one of their existing GRX IP addresses for the GTP Hub they sponsor and *include the APNs of the clients of the GTP Hub in their local DNS pointing to the GTP Hub IP*. It does not matter if several network clients use the same APN (NI) such as "internet". This is the "tunnel method" below.

Method (a) is technically *simple, but the effective opening is slow* as all the potential VPLMNs will take several months to process the new sponsors IR21 which include the GTP Hub IP. An example is the Telecom North America (TELNA) experience which has multiple GRX sponsors some with method (a) and others with (b), for their multi-IMSI service. One partner (that shows that few are checking) reached with (a)-type sponsors, recognized that an IP address in the IR21 belonged to Telna and questioned the sponsor. Also, the roaming data service worked in many countries, it was failing from VPLMNs which had not yet added Telna's IP address after several months.

3.3.2   Sponsor-RH IP Address Method Using a Tunnel between the Sponsor and the GTP Hub

The second method, shown in Figure 3.1, will be immediately operational as it does not involve work at the VPLMNs.

The first step is that *the RH must obtain from each of his sponsors one of their normal GRX IP address* which we call "Sponsor-GTPH" IP address. As we assume that there is a GPRS roaming agreement between the Sponsor and the VPLMN, messages with this Sponsor-RH IP address from/to the VPLMN are not barred by the VPLMN's firewall and reach the concerned sponsor's GGSN. It is not possible because of the GRX ASN

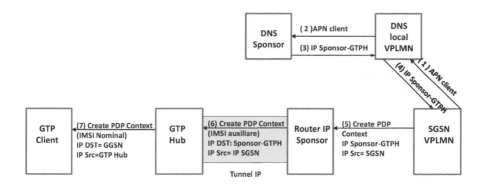

Figure 3.1 IP Tunnel between a sponsor's IP network and the GTP Hub.

routing that the GRX could be configured to reach the GTP Hub directly. Between the sponsor and the GTP Hub operator, *there is a transparent IP tunnel.* The IP messages from the VPLMN to the Sponsor-GTPH IP are relayed by the sponsor to the GTP Hub (the source IP (VPLMN) and the destination IP (Sponsor-GTPH) will be unchanged. The GTP Hub will respond on the same tunnel and the sponsor's network will relay transparently to the VPLMN.

This means that the GTP Hub has several IP addresses configured, his own and all the "Sponsor-GTPH", that it listens.

When it responds to a received Create PDP Context Request (the VPLMN is a classical 3G) or to a Create Session Request (GTPv2 of LTE), it will just set the GGSN address control plane and the GGSN address data *plane to the Sponsor-GTPH in the IP destination* value in the received message. From the GTP version received, it knows if it is LTE and the response will be a Create Session response with also the Sponsor-GTPH in the "Fully qualified TEID" parameters concerning the Control Plane and the User Plane.

3.4 Transparent and Non Transparent Parameters Relayed by a GTP Hub

A GTP Hub *is not transparent* for the GTP-C protocol *as an IP router is* for the visitor's data traffic. For a multi-IMSI service, the used IMSI auxiliary (any) is replaced by IMSI nominal. The trace of the Create PDP Context received by the GTP Hub and the one that it sends shows it.

_____ Header IP received _____
  Src IP '192.168.0.16'
  Dst IP '192.168.0.4'
- - - - Super Detailed GTP, SIP, DIAMETER, RADIUS Analyser (C)HALYS - RS=1 - -
    (32)GTP version 1, ProtocolType=(1):GTP-C, T(v2) ou Spare=0, E=0, S=1, PN=0
    Message Type:(16):Create PDP Context Request
    Length_Payload 106
    TEID 00000000
    SequenceNumber 10241
    N-PDU 00
    Next_extension_header_type 00

    _____
       IMSI: 340990100000005 /* AUXILIARY IMSI ! */
       Recovery:Restart_counter=74 /* of the SGSN */
       TEID Data I: 00000001
       TEID Control Plane: 00000001
       NSAPI 0
       Charging Characteristics: 0800
       Charging Characteristics: 0800
       APN(Access Point Name): netgprs.com
       Protocol Configuration Options: 001580C0231101010011036D69670868656D6D656C
       SGSN Address for Control Plane: 192.168.0.16
       SGSN Address for user traffic: 192.168.0.16
       MSISDN 46702123456
       QoS profile: 000B921F
         /* GTP Hub processing */
           INFO :: Context 0 : APN breakthrough = 0
           INFO :: Context saved : TEIDcp_sgsn = 1, TEID_data_sgsn = 1
           IMSI nominal found : 208104286725783 (IMSI auxiliary = 340990100000005)
           /* prepare for DNS access by the GGSN behaving as a SGSN */
           Operator Identifier of IMSI nominal 208104286725783 : mnc010.mcc208.gprs
           APN netgprs.com does not have an Operator Identifier, we add it...
           new APN : netgprs.com.mnc010.mcc208.gprs /* APN=OI-NI will be used for DNS
    access */
           Looking for GGSN IP address for APN netgprs.com.mnc010.mcc208.gprs
           APN : netgprs.com.mnc010.mcc208.gprs => IP address : 192.168.0.159, priority = 1
_____ Header IP sent _____
  Src IP '192.168.0.4'
  Dst IP '192.168.0.159'
- - - - Super Detailed GTP, SIP, DIAMETER, RADIUS Analyser (C)HALYS - RS=0 - -
    (32)GTP version 1, ProtocolType=(1):GTP-C, T(v2) ou Spare=0, E=0, S=1, PN=0
    Message Type:(16):Create PDP Context Request
    Length_Payload 106
    TEID 00000000
    SequenceNumber 10241
    N-PDU 00
    Next_extension_header_type 00

    _____
          IMSI: 208104286725783 /* NOMINAL IMSI now ! */
          Recovery:Restart_counter=1 /* changed to GTPHUB counter !! */
          TEID Data I: 00000001   /* Transparent */
          TEID Control Plane: 00000001   /* Transparent */
          NSAPI 0   /* Transparent */
          Charging Characteristics: 0800 /* Transparent */
          Charging Characteristics: 0800 /* Transparent */
          APN(Access Point Name): netgprs.com /* Transparent */

Protocol Configuration Options:
001580C0231101010011036D69670868656D6D656C
       SGSN Address for Control Plane: 192.168.0.4 /* changed to GTP Hub IP control
plane */
       SGSN Address for user traffic: 192.168.0.4 /* changed to GTP Hub IP user
plane */
       MSISDN 46702123456   /* Transparent */
       QoS profile: 000B921F /* Transparent */

We also see that the Restart Counter is changed to be the number of restarts of the GTP Hub and that when the two SGSN addresses are replaced by those of the GTP Hub, the other parameters are transparent.

3.5 Transparent Tunneling of the User Data (GTP-U)

We have seen in Figure 2.1 that the GTP Hub behaves as a SGSN which uses the home GGSN and has obtained the IP address from the GGSN pool, which it uses transparently. It returns the same IP address to the mobile, which is the EndUserAddress returned to the SGSN, when a GTP Hub is used instead of a classical GGSN.

This is shown in Figure 3.1.

So the response from airfrance.fr will be sent back to the HPLMN GGSN because this address belongs to the GGSN range in the home PLMN.

3.6 Local Break-Out (LBO) Static and Dynamic [3.1]

3.6.1 Definition: Static and Dynamic LBO

The LBO is "static" when the IP address allocated to the mobile is given by the GTP Hub or the GGSN. This is the case of the EU BEREC scheme below.

It is "dynamic" when the IP address is still allocated by the HPLMN GGSN but dynamically changed by the GTP Hub for certain APN, websites or service PORTs (such as SIP) as subscribed by a visitor. The change is only in the GTP-U encapsulated IP address. For other APNs, websites or service PORTs of this visitor, there is no LBO, so that, for example, the MMS service which must be from the HPLMN remains available.

3.6.2   EU BEREC Proposed Technical Scheme for LBO

BEREC assumes that the visitor has changed his internet APN which defines the path used to find a GGSN to access the internet. This involves, for each visitor, changing the default internet setting from their HPLMN internet setting to EUinternet. The scheme assumes that the standard VPLMN GGSN is used.

Even if the VPLMN may help them with an OTA GPRS service for OTAble handsets (settings may depend on the telephone model), it will not take care of restoring the default internet when the visitor is back home. This makes the BEREC scheme unacceptable for most visitors. However, for the rare case where visitors will use it, it is straightforward and they establish the data sessions (Create PDP Context) with the VPLMN GGSN without any use of their HPLMN GGSN. When they have subscribed to the LBO, their MSISDN and a credit of X Mega octets are entered in the VPLMN RADIUS server. When they have used their credit, the data service through LBO is interrupted. Again, all this is straightforward with standard GGSN capabilities. Note that provisioning based on IMSI requires more development as it usually requires use of the Welcome SMS information or the VLR data to determine the IMSI from the MSISDN. With LBO, the only provisioning that the VPLMN has to do in its network is:

EUinternet.*mncXXX.mccYYY.gprs* -> IP address of the VPLMN GGSN in the local DNS used by its SGSNs.

It must help the visitors to activate the APN EUinternet as default internet APN, even if it was provisioned by their home operator. Note that there is an OMA standard [3.5] which would provision the handset from the SIM card but no terminal implements it. To date, there is a file in the SIM cards called "EFapn" which contains a list of authorized APNs controlled by the operator, EUinternet should be included in it. So the VPLMN would need a classical OTA server to activate the break-out APN with the difficulty of different models, iPhones and Blackberrys which do not follow the OMA standard. This makes the system below very attractive because it does not need any change of APN.

3.6.3 LBO Service Provisioning in the VPLMN without Any APN Change Using a GTP Hub

This alternative implementation of LBO assumes for simplicity that the VPLMNs have a GTP Hub-based GGSN *for all the visitors' data traffic,* whether they subscribe the LBO service or not. The principle was given by Figure 2.3. But this GTP Hub and the charging could also be provided by hosting with a third party.

The major difference between a GTP Hub-based GGSN architecture and a classical GGSN is that the Create PDP Context is relayed *but can be processed* differently to perform an interrogation of the RADIUS server using the MSISDN to know if the LBO service has been subscribed to. A GGSN does not interrogate RADIUS for visitors. There is also a *difference of configuration* between a GTP Hub bi-IMSI of section 2.1.1 and a GTP hub for LBO. The *IMSI auxiliary<->IMSI nominal table is made transparent*, the table is void and the GTP Hub LBO uses the same IMSI it has received from the SGSN to create a PDP Context with the HPLMN GGSN.

The solution is that the Create PDP Context is still established with the HPLMN so there is no need to change the APN to use the VPLMN GGSN as in the EU BEREC scheme. But in the case of LBO, there will not be any user traffic with the HPLMN GGSN (and charging), the user traffic will go through the VPLMN GGSN. This is shown in Figure 3.2.

In the SGSN local DNS, all APN WEBs of visitors of the countries which may have LBO are set to the IP address of the GTP Hub. Remember that an APN includes two parts OI and NI (mcc010.mnc208.gprs) as below:

web.sfr.mnc010.mcc208.gprs   ->IP address of the VPLMN GTP Hub
orange.fr.mnc001.mcc208.gprs   ->IP address of the VPLMN GTP Hub
a2bouygtel.mnc020.mcc208.gprs -> IP address of the VPLMN GTP Hub
free.mnc015.mcc208.gprs   ->IP address of the VPLMN GTP Hub

[3.1]

This is hardly more complicated for the VPLMN than the BEREC scheme as the list of APN WEBs is found in the roaming partners' IR21 in the standard section APN WEB LIST, it is shown as action (1) in Figure 3.2.

On subscription to the LBO service, the MSISDN of the visitor (this is known and there is no need to know the IMSI), is added in the white list of the VPLMN RADIUS server with the credit: action (2) in Figure 3.2.

In the case of an MSISDN (or IMSI) or a particular APN belonging to the white list, there is what we call a "static local break-out", the GGSN behaves as a GGSN for this APN for the user traffic which means that when the Create PDP Context from the SGSN is received, the EndUserAdress is allocated from the GTP Hub pool. And airfrance.fr will send responses to the GGSN using as destination the "IP allocated by Hub" type End User Address allocated to the mobile.

For dynamic local break-out, it is a combination of "Transparent" and "static local break-out". The EndUserAddress allocated to the mobile is still that of the GGSN, like for non-break-out. All encapsulated packets received from the mobile will have this source address. This is because some URLs may be without or with break-out within the same session.

But an "IP allocated by Hub" has been also allocated in the PDP Context to the mobile which will be used for an eventual URL with dynamic break-out.

For such a URL, when the HTTP request is sent directly to airfrance.fr, the encapsulated IP source is set to "IP allocated by Hub" by the GTP Hub. When the response arrives from the web site, it is changed by the GTP Hub to "IP allocated by GGSN" which has been allocated to the handset by the Create PDP response. This is different from the normal behavior of a GGSN which is always transparent for the encapsulated IP addresses.

3.6.4   Details of the LBO Service Implementation by the VPLMN Using a GTP Hub

This is represented in Figure 3.2 and shows the different cases. In the detailed chapter on RADIUS Hubs, we can see that MSISN, IMSI, APN and PORT may be part of the RADIUS Access-Request sent to the RADIUS server to know if this MSISDN is authorized for LBO. The direct Internet interface is called Gi.

3.6.4.1   An Alternative Dynamic Implementation of the EU BEREC Scheme

The list of MSISDN and the corresponding APN Internet (OI) has been entered in the RADIUS server with a data volume credit. When the Create PDP Context is received from the SGSN, because of the DNS setting [3.1], it is relayed to the HPLMN GGSN, but an EndUserAddressHub is also assigned for LBO by the GTP Hub from his "Pool IP Hub", while the EndUserAddress allocated by the GGSN is assigned to the mobile. It is better than the static implementation which would allocate the EndUserAddressHub — another APN (MMS) which does not use LBO —

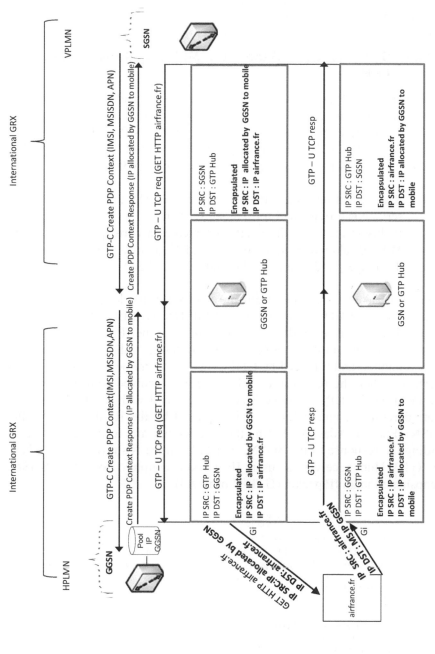

Figure 3.2 Transparent relay of HTTP traffic.

this IP would not work with the HPLMN GGSN. This way both LBO and non-LBO APNs may be used by the visitor.

For non-LBO APNs such as mms.sfr.fr.mnc010.mcc208.gprs, the VPLMN uses the normal GGSN IP address.

3.6.4.2 Dynamic LBO Based on MSISDN Subscription and APN

As illustrated by Figures 3.3 and 3.4, the MSISDN and the internet APN such as orange.fr are entered (2) by the sales service of the VPLMN in the RADIUS server. If VoiP is allowed in LBO, the SIP port 5060 is also entered. When the PDP Context Response is received (3) by the GTP Hub with the EndUserAddress (which is the IP allocated by the GGSN):

- RADIUS is interrogated (4) with the MSISDN to know if the LBO service is provisioned for that subscriber.
- If this is the case, the GTP Hub creates a correspondence in the MAPIP table with its own allocated IP address. The existence of this correspondence is used to indicate the LBO service.

Then, if a packet is received from the SGSN (on the left), the GTP Hub uses the EndUserAddress to find if the MSISDN has LBO. If this is so, the IP SRC is replaced by EndUserAddressHub. But the IP and TCP or UDP checksum depend on the IP SRC and are no longer valid. The checksums are recomputed and replaced by CHECKSUM before being sent on the Gi interface. When the response (a WEB page) is received from internet (Gi interface) on the left, the table MAIP is searched with the IP DST (End UserAddressHub) and if found, it is replaced by the IP allocated by the GGSN (EndUserAdress). Again, the checksums are recomputed before sending the response to the SGSN. A detailed trace is shown below where the GTP Hub receives a user IP packet coming from the SGSN, encapsulated in an IP packet (GTP-U protocol).

```
_____ Header IP received _____
Src IP '192.168.0.16'    /* Address of SGSN
Dst IP '192.168.0.4'     /* Address of GTP Hub
-- - Super Detailed GTP, Gi, SIP, DIAMETER, RADIUS Analyser (C)HALYS - RS=1 ENCAPS=1 - -
          (32)GTP version 1, ProtocolType=(1):GTP-C, T(v2) ou Spare=0, E=0, S=1, PN=0
          Message Type:(255):G-PDU(IP packet tunneled)
          Length_Payload 280
          TEID 00000006

          _____
          IP Header starts at iS = 12: Version 4 Header_length = 20
          IP Protocol = (6):TCP
          Header IP checksum in the header: AB01
```

Header IP checksum calculated: AB01
IP_Src: **192.168.0.52** IP_Dst: 108.174.146.20 (IP of airfrance.fr)
  TCP length (from IP envelop) 256
  TCP checksum in the header: AE00
  TCP checksum calculated: AE00
   payload Length = 224

  Port HTTP
   GET / HTTP/1.0 Host: www.airfrance.fr Accept: text/html, text/plain, text/css,
text/sgml, */*;q=0.01 Accept-Encoding: gzip, bzip2 Accept-Language: en User-Agent:
Lynx/2.8.7rel.1 libwww-FM/2.14 SSL-MM/1.4.1 OpenSSL/1.0.0j

The GTP Hub extracts the encapsulated packet IP and analyzes the destination IP. If it belongs (198.174.146.20 = airfrance.fr) to TSTIPEIP, the source IP EndUserAddress 192.168.0.52 is replaced by EndUserAddressHub 192.168.0.183, a local address allocated by the GTP Hub. The table MAPIP created when the Create PDP Context is performed is used in order to provide the LBO for such websites as www.airfrance.fr or www.telna.com

  EndUserAddress<->EndUserAddressHub

 But the IP Header and the TCP or UDP Header contains "checksums" which both depend in their computation on the IP Source. These checksums (standard function of the OS) are for data integrity. If the same checksum was kept, the message would be rejected by the network. In our scheme, the checksums are recomputed after the change of source IP as shown in the trace below of the message sent in LBO on the Gi interface:

http://www.airfrance.fr/

  - - Super Detailed GTP, Gi, SIP, DIAMETER, RADIUS Analyser (C)HALYS - RS=0 ENCAPS=0 -
  -

   ---- SENT   Gi Interface trace GGSN-GTP Hub -> Internet
   IP Header starts at iS = 0: Version 4 Header_length = 20
   IP Protocol = (6):TCP
   Header IP checksum in the header: AB01
   Header IP checksum calculated: **AA7E**   /* recomputed checksum
   IP_Src: **192.168.0.183** IP_Dst: 108.174.146.20
     TCP length (from IP envelop) 256
     TCP checksum in the header: AE00
     TCP checksum calculated: **AD7D**   /* recomputed checksum
      payload Length = 224

   Port HTTP
    GET / HTTP/1.0 Host: www.airfrance.fr Accept: text/html, text/plain, text/css,
text/sgml, */*;q=0.01 Accept-Encoding: gzip, bzip2 Accept-Language: en User-Agent:
Lynx/2.8.7rel.1 libwww-FM/2.14 SSL-MM/1.4.1 OpenSSL/1.0.0j

The IP checksum received was AB01 before and is recomputed as **AA7E**. The TCP checksum received was AE00 and is **AD7D** after recomputation.

The website answer on the Gi interface, the destination IP is 192.168.0.183 which had been allocated by the GTP Hub. Here is the packet:

```
- - - Super Detailed GTP, Gi, SIP, DIAMETER, RADIUS Analyser (C)HALYS - RS=1 ENCAPS=0 - -
        ---- RECEIVED Gi Interface trace GGSN-GTP Hub <- Internet
        IP Header starts at iS = 0: Version 4 Header_length = 20
        IP Protocol = (6):TCP
        Header IP checksum in the header: 7DD4
        Header IP checksum calculated:   7DD4
        IP_Src: 108.174.146.20 IP_Dst: 192.168.0.183
          TCP length (from IP envelop) 1416
          TCP checksum in the header: 7999
          TCP checksum calculated:   7999
            payload Length = 1384
```

```
        Port HTTP
            HTTP/1.1 200 OK  Server: nginx/1.2.1  Date: Fri, 21 Dec 2012 16:12:07 GMT
Content-Type: text/html;charset=utf-8  Connection: close  Status: 200  X-Powered-By: Phusion
Passenger (mod_rails/mod_rack) 3.0.18  X-Frame-Options: sameorigin  X-XSS-Protection: 1;
mode=block  Content-Encoding: gzip
```

Using the MAPIP table, the GTP Hub changes the IP Destination address and replaces it by 192.168.0.52.

This response is encapsulated in a GTP-U IP packet. The checksums (7DD4 for IP before, **7E57** after) are recomputed, and sent to the mobile through the tunnel with the SGSN.

```
Header IP sent _____
        Src IP '192.168.0.4'          /* Address of GTP Hub
        Dst IP '192.168.0.16'         /* Address of SGSN
-- - Super Detailed GTP, Gi, SIP, DIAMETER, RADIUS Analyser (C)HALYS - RS=0 ENCAPS=1 - -
        (32)GTP version 1, ProtocolType=(1):GTP-C, T(v2) ou Spare=0, E=0, S=1, PN=0
        Message Type:(255):G-PDU(IP packet tunneled)
        Length_Payload 1440
        TEID 00000001
```

```
        IP Header starts at iS = 12: Version 4 Header_length = 20
        IP Protocol = (6):TCP
        Header IP checksum in the header: 7DD4
        Header IP checksum calculated:   7E57 /* recomputed checksum
        IP_Src: 108.174.146.20 IP_Dst: 192.168.0.52
          TCP length (from IP envelop) 1416
          TCP checksum in the header: 7999
          TCP checksum calculated:   7A1C        /* recomputed checksum
            payload Length = 1384
```

Port HTTP
HTTP/1.1 200 OK Server: nginx/1.2.1 Date: Fri, 21 Dec 2012 16:12:07 GMT Content-Type: text/html;charset=utf-8 Connection: close Status: 200 X-Powered-By: Phusion Passenger (mod_rails/mod_rack) 3.0.18 X-Frame-Options: sameorigin X-XSS-Protection: 1; mode=block Content-Encoding: gzip

3.6.4.3 Dynamic LBO Based on Website and MSISDN Subscription

This is a more restricted service (cheaper or free) for a limited number of domain names, for example, the help desk of the VPLMN. In this case, there is a provisioning interface directly on the GTP Hub to create the list of LBO domain for the MSISDN. When a browser wants to access a domain such as airfrance.fr, yahoo.com, it makes a DNS interrogation and obtains a list of IP addresses. It selects one of them and will use it as the IP DST address for the HTTP GET request of a page. The DNS request is part of the user data and is sent to the HPLMN GGSN (an optimization could make that LBO also). Here is an example of the DNS response received from the HPLMN GGSN.

_____ Header IP received _____
    Src IP '192.168.0.159'   /* address of HPLMN GGSN
    Dst IP '192.168.0.4'   /* address of GTP Hub
  - - -Super Detailed GTP, Gi, SIP, DIAMETER, RADIUS Analyser (C)HALYS - RS=1 ENCAPS=1 -
-

    (32)GTP version 1, ProtocolType=(1):GTP-C, T(v2) ou Spare=0, E=0, S=1, PN=0
    Message Type:(255):G-PDU(IP packet tunneled)
    Length_Payload 111
    TEID 00000001
    SequenceNumber 0
    N-PDU 00
    Next_extension_header_type 00

    _____
    IP Header demarre a iS = 12: Version 4 Header_length = 20
    IP Protocol = (17):UDP
    Header IP checksum in the header: 7165
    Header IP checksum calculated: 7165
    IP_Src: 212.27.40.240 IP_Dst: **192.168.0.52**
      UDP Header source_Port 53 dest_Port 54133
      Total UDP length header+data (in UDP header) 87
      UDP checksum in the header: 06AC
      UDP checksum calculated: 06AC
       payload Length = 79

    _____
    payload IP
    Port Domain Name System
     Transaction ID 981C
     Flags   8180
      DNS QR: (1):response
      DNS Operation Code: (0):request
      Questions Resource Record Count: 1

Answer Resource Record Count(number of IP addresses) 3
Authority Resource Record Count 0
Additional Resource Record Count 0
domain: www.google.fr
Query Type Host Address (0): standard
Query Class 01C0
Type Host Address (1):A (IPv4)
Class 0100
Time to live: 883 seconds
IP Address Length: 4
IP Addr: 173.194.34.24
Type Host Address (1):A (IPv4)
Class 0100
Time to live: 883 seconds
IP Address Length: 4
IP Addr: 173.194.34.31
Type Host Address (1):A (IPv4)
Class 0100
Time to live: 883 seconds
IP Address Length: 4
IP Addr: 173.194.34.23

To implement the LBO, the GTP Hub analyzes all DNS responses and creates a list TSITEIP showing which IP address corresponds to which domain name. The system has a simple DPI module (Deep Packet Inspection) in Figure 3.3 which selects DNS responses (PORT=53) and

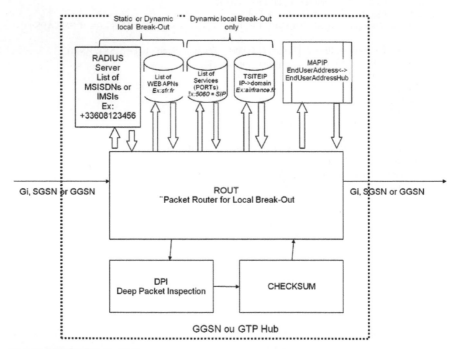

Figure 3.3 GTP Hub for LBO.

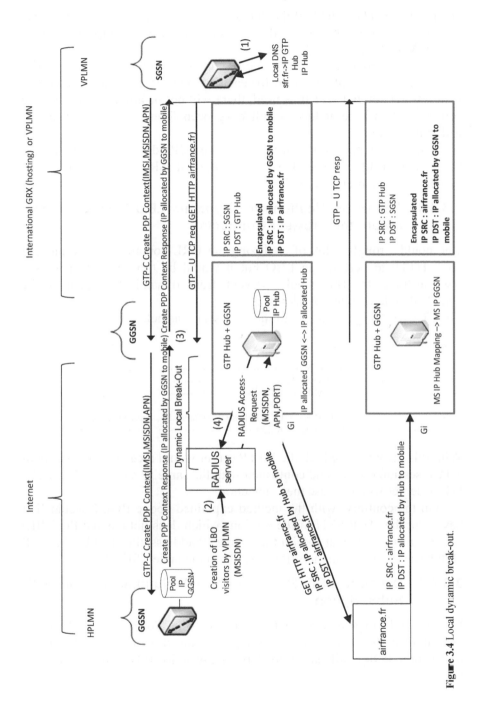

Figure 3.4 Local dynamic break-out.

updates the TSITEIP table if the domain name, for example *www.google.fr* above, is one of the privileged domain for which LBO is implemented for certain MSISDNs.

But the browser is "caching" the IP addresses and does not necessarily make a new DNS interrogation for each website. If a "cache IP address" which is not yet in TSITEIP is used, there will not be any LBO. But after a certain time, all the addresses will have been found as explained in the following section.

Whenever a packet is sent to one of these IP addresses, if the SRC IP belongs to a LBO subscriber, the change of SRC IP takes place as explained in the previous section.

3.6.4.4   The IP Destinations for Dynamic LBO Must be known before the Browser Connects

One could wrongly think of analyzing the URL contained in the internet GET HTTP to perform an LBO but it would be too late, when we see www.airfrance.fr. This would prevent from dynamically updating the TSITEIP table with the IP addresses.

The TCP session (connection of the browser) is as below:

```
Server HTTP                                                          Browser

<----------------------------TCP SYN ----------------------------------   connects the TCP session
----------------------------------TCP SYN+ACK ----------------------------->
<----------------------------TCP ACK ----------------------------------   confirms the TCP session

<----------------------------TCP ACK + GET HTTP www.airfrance.fr--   ask for a page
----------------------------------TCP ACK + OK 200 ------------------------>   receives the page
```

With this incorrect method, the HTTP server would receive a GET without a TCP session opened and confirmed with the same IP source address. The HTTP server would refuse the connection.

On the contrary, with the method explained above ROUT would first have sent the TCP SYN of the session (which does not contain the URL) *already using LBO* that is with the source IP address allocated by the GTP Hub because the destination IP is already in the table TSITEIP.

3.6.4.5   Dynamic LBO Based on PORT Number and MSISDN Subscription

Is it possible to subscribe to LBO just for the VoIP service? Yes, the procedure is the same as above except that the trigger to start LBO is a destination PORT SIP subscribed by a given MSISDN in the "List of Services".

The VPLMN could obviously include PORT 53, which is used for the DNS service, among the PORT list of all visiting subscribers avoiding as in the trace above to have all these requests sent to the HPLMN while they could be answered by the local DNS. There are side effects as certain websites might be language dependent based on the source IP.

3.6.5 Computation of the Average Number of DNS Draws to Have All the IP Addresses

To quickly enable the LBO to a specific domain name, the list of IP addresses must be as comprehensive as possible (Google has about 100 and gives only a few in each DNS response). How quickly is the list completed? Until the list is complete, some internet access will not use LBO although the visitor has subscribed the service.

We note as M the total number of IP addresses for a given domain name such as www.google.fr and T the number of different sets of IP addresses that the DNS returns for www.google.fr. This is assumed to be a known fixed number which depends on the domain. The problem (the PANINI album) is then exactly the same as boys in the 1980s were facing when they tried to fill a free picture book of soccer players ($M = 220$), but they had to buy many envelops with $T=10$ different players which they might already have. How many envelops do they expect to buy to fill the album (if there was not the obvious possibility to exchange with their comrades)?

Once this simple problem is resolved, we will address the case (like the DNS search case) when M is not known but follows the known law of probability where the average and the mean are known.

3.6.5.1 Conditional Probability of k New IP Addresses for Each New Draw

Let $\Pi\{k \mid q\}$ be the probability of k new IP addresses, if q different are already known. Each draw is T different new IP addresses and their total (assume it is known) is T). Note   the number of already known IP addresses for this domain.

Note:

A_N^k is the number of arrangements (the order counts) of k different items drawn from N, which is

$$A_N^k = \frac{N!}{(N-k)!}$$

A_N^k is the number of combinations (the order does not count) of k different items drawn from N, which are, 123, 132, 213... All six combinations are considered the same:

$$C_N^k = \frac{N!}{(N-k)!k!}$$

with as usual

$$N! = (N(N-1)(N-2)\ldots 2.1$$

One computes using conditional probabilities (example of a draw of T=6 with k=2 news (N) and T−k=4 old(O) already found: O N O N O O):

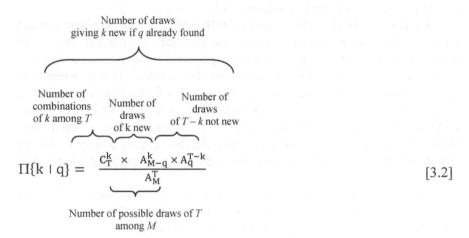

$$\Pi\{k \mid q\} = \frac{C_T^k \times A_{M-q}^k \times A_q^{T-k}}{A_M^T}$$ [3.2]

The numerical result of the $\Pi\{k \mid q\}$ is given below for $M = 12$ different IPs and each DNS query is assumed to return T=3 different IP addresses (but for each draw, a previous address could appear). One can check that $\Pi(3 \mid 0) = 1$, that is, if q= 0 (no IP address has yet be found), it is certain that k=3 new addresses will be obtained. Also $\Pi\{0 \mid 12\} = 1$, that is, if all IP addresses have been found, it is certain that $k = 0$, no new addresses can be found. The last column sums the probabilities for all possible values of k new IP addresses, it is always 1.

If $q = 11$, there is a **0.25** chance of finding the last one which is intuitive with T=3 and M=12 (T/M=0.25).

We note $P_i(q)$ is the probability of having found q addresses on the ith draw.

The event "q addresses known" may be obtained from the events q, q-1, q-2 at the ith draw, with a draw of 0, 1, 2, ...T. The rule of conditional probabilities use the previous function $P\{k \mid q\}$ and yields:

$$P_i(q) = {}_iP_{i-1}(q) \times \Pi\{0 \mid q\} + P_{i-1}(q-1) \times \Pi\{1 \mid q\}$$
$$+ P_{i-1}(q-2) \times \Pi\{2 \mid q\} + \cdots + P_{i-1}(q-T) \times \Pi\{T \mid q\}$$

We initialize with :

$$P_0(0) = {}_i1, P_0(1) = P_0(2) = \ldots = P_0(M) = 0$$

For the above example, you have the results for up to 20 draws of $T=3$ IP addresses as shown in Table 3.1 and 3.2.

Table 3.1 Table of $\Pi\{k \mid q\}$ for M=12 to be found and T = 3 draws

New addresses

$\Pi\{k \mid q\}$	0	1	2	3	Total probability
0	0.00000	0.00000	0.00000	1.00000	1.00000
1	0.00000	0.00000	0.25000	0.75000	1.00000
2	0.00000	0.04545	0.40909	0.54545	1.00000
3	0.00455	0.12273	0.49091	0.38182	1.00000
4	0.01818	0.21818	0.50909	0.25455	1.00000
5	0.04545	0.31818	0.47727	0.15909	1.00000
6	0.09091	0.40909	0.40909	0.09091	1.00000
7	0.15909	0.47727	0.31818	0.04545	1.00000
8	0.25455	0.50909	0.21818	0.01818	1.00000
9	0.38182	0.49091	0.12273	0.00455	1.00000
10	0.54545	0.40909	0.04545	0.00000	1.00000
11	0.75000	0.25000	0.00000	0.00000	1.00000
12	1.00000	0.00000	0.00000	0.00000	1.00000

If q addresses already found

Table 3.2 Table of $P_i(q)$ for $M=12$ and $T=3$

$P_i(q)$	0	1	2	3	4	5	6
1	0.0000	0.0000	0.0000	1.0000	0.0000	0.0000	0.0000
2	0.0000	0.0000	0.0000	0.0045	0.1227	0.4909	0.3818
....							
10	0.0000	0.0000	0.0000	0.0000	0.0000	0.0000	0.0000
....							
20	0.0000	0.0000	0.0000	0.0000	0.0000	0.0000	0.0000

$P_i(q)$	7	8	9	10	11	12	mean
1	0.0000	0.0000	0.0000	0.0000	0.0000	0.0000	3.0000
2	0.0000	0.0000	0.0000	0.0000	0.0000	0.0000	5.2500
....							
10	0.0000	0.0005	0.0123	0.1137	0.4093	0.4642	11.3242
....							
20	0.0000	0.0000	0.0000	0.0004	0.0373	0.9623	11.9619

Table 3.3 Estimate of the number of draws to terminate various samplings experiments

Type of "game"	M	T	Estimate of number of draws to reach stop value M-0.5	Stop value
The PANINI album	220	10 pictures	131	219.5
Google DNS	100 (estimated)	3 IPs (observed)	174	99.5

Each column $0 <= q <= M$ gives the value $P_i(q)$ for each of the draws 1 to 20. On the 20th, you see that the probability of having found all the 12 IP addresses reaches 0.9623 with a mean value of 11.9619 (close to the real total value 12).

One sees that we have $P_i(q <= 3) = 0$ for any i, because we draw $T=3$ IP addresses, so we have three IPs already on the first draw.

The last column gives for draw # i the average number of IP addresses found using the usual definition of the mean:

$$\text{mean} = \Sigma_i(i * P_i(q)) \tag{3.3}$$

3.6.5.2   Numerical Results for Certain Specific Problems

We want to compute the estimate of the number of draws such that mean >
M-0.5 (we are close to the total number) using [3.2] (note that the C
program uses double precision floating numbers because they are very
large)

```
/*-----------------------------------------------------------------------*/
/* FUNCTION: conditional probability PI(k | q) */
/*-----------------------------------------------------------------------*/
double PI(int k, int q)
{
        return   ((CNk(T,k)*ANk(M-q,k)*ANk(q,T-k))/ANk(M,T));   //formula  3.2   of
Chapter 3
}
/*-------- Estimate of the number of draws necessary to find q different samples -----*/
/* the PANINI soccer player game                                           */
/* A.Henry-Labordere dec 2012                                              */
/*-----------------------------------------------------------------------*/
int main()
{
double P = 0.0; //cumulated probability for all k
long i;
long q;
int draws; // number of draws
double mean; // mean of number found after draw  # i
double PIim[M+1][T+1];
double Pim1[M+1];   //probability  to have found q at i-1 th draw
double Pi[M+1];       //probability to have found q at ith draw
        //1) computes the PI(q|k) table
        for (q = 0; q <= M; q++)
          Pi[q] = Pim1[q] = 0.0; // init to 0 initially
        Pim1[0] = 1.0; //init probability  0 found = 1, proba >= 1 found = 0;
        printf("---- T= %d different draws among M %d\n",T,M);
        printf("q\\k");
        for (i = 0; i <= T ; i++)
           printf("   %2d  ",i);
        printf("  Total probability\n");
        for (q = 0; q <= M; q++)
        {
          printf(" %2d ",q);
          P = 0.0;
          for (i = 0; i <= T ; i++)
          {
            printf(" %6.4f  ",PIim[q][i]=PI(i,q)); // PI according to formula [3.2]
            P += PI(i,q);
          }
          printf("   %6.4f \n",P);
        }
        //2) computes   Pi(q)
```

```
            printf("---- Display all Pi(q)\n");
            printf("q");
            for (q = 0; q <= M; q++)
             printf("   %d   ",q); //
            printf("  mean \n");
  for (draws = 1; draws <= DRAWS; draws++)
  {
            for (q = 0; q <= M; q++)
            {
             Pi[q] = 0.0;

             for (i = 0; (i <= T) && (i <= q); i++) // 0<= i <= T
               {
                Pi[q] += Pim1[q-i] * Plim[q-i][i];
               } // end for  i
              printf ("%1.4f ",Pi[q]);
            } //end for q
            for (i = 0; i <= M; i++)
            {
             Pim1[i] = Pi[i]; // sets Pi-1 probabilities at draw i-1
            }
            //computes mean
            mean = 0.0;
            for (i = 0; i <= M; i++)
             mean += i * Pi[i];  //according to formula [3.3]
            printf("    %1.4f\n", mean);
            //stoping rule
            if (mean > M-0.5) //if mean very close to known number M
            {
                       printf("stop buying: average %d envelops purchased\n",draws);
                       break;
            }

    } //end for draw
}
```

3.7   GTP Hubs for LTE 4G

3.7.1   LTE Network Architecture and Call Flows of a PGW

We turn to the case of GTP Hubs for LTE networks. It is necessary to explain the System Evolution Architecture (SAE), and to detail the PDN Gateway (PGW) which has the same role as the GGSN, as well as the protocols which a LTE GTP Hub should support with the support of call flows.

The System Architecture Evolution (SAE) provides a new Evolved Packet Core which replaces the previous 3G Packet Core based on SGSNs and GGSNs. It is illustrated by Figure 3.5. The four main evolutions are:

- The User Traffic (UP or User Plane) handling is handled by a separate dedicated equipment called SGW (Serving Gateway). The MMEs which replace the SGSNs *only handle the Control Traffic (CP or Control Plane)* with an evolution v2 of the GTP protocol which allows to manage the quality of bearer services. The main advantage is that when the traffic grows, only the SGW needs to be upgraded.

- The SGW *is not just a relay for the Control Plane signalling traffic*, as many call flows on internet assume. *It adds IP addresses concerning the data User Plane* (he knows, not the MME) for messages from the MME to the PGW and suppresses some Control Plan IP address from the PGW to the MME.

- A Policy, Charging and Control system is part of the SAE. It allows to individually adjust the service quality to the applications need and the users' *individual subscriptions*. The PCEF (DPI is an almost mandatory part of it), is integrated in the standard PGW model, although it could be technically separated and inserted on the SGi (Gi in 3G) interface with probes on the S5/S8 (Gn in 3G) interface to capture the various session parameters in the Create Session Request and Response messages. *There is no RADIUS server in 4G,* the authentication function is performed by the PCRF (Gx/DIAMETER protocol) and the UE IP address allocation is mostly performed by an internal IP pool in the PGW. Even in 3G, a RADIUS is a legacy heritage with little use as the authentication and accounting are performed by the PCRF.

- The User Equipment is always connected, not like in 3G with an "Attach phase" and a separate session creation (establishment of a PDP context) when the user wants to connect to a service. In LTE, when the user "attaches" a default session, he is automatically created with a default Quality of Service (the default "Bearer Context"). As the users must always remain connected even when not using the access to network services, the tunnels are maintained by the SGW when the users are idling their data traffic. This allows to reconnect the users at the request of external services such as IMS (push data). For example, it makes it possible to receive calls with VoIP/LTE.

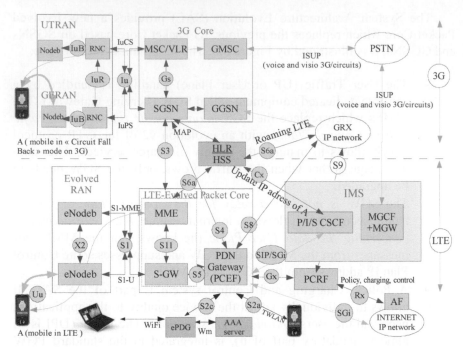

Figure 3.5 LTE network associated with a 3G network.

The first two evolutions exist in the recent 3G data networks architectures having PDN Gateways with PCEF and PCRF. For medium size networks, the recently combined PGW-GGSNs provide 3G and the 4G data services simultaneously. It is easy from the GTP version indication in the GPT messages to know if it is 3G GTPv1 or LTE GTPv2 in the same software. Also, the SGW and the *PGW may use the same hardware although these are distinct functions* with different IP addresses in the same equipment. The protocol GTP-U for the data traffic is the same as in 2.5G or 3G.

To understand the call flows in the roaming cases below, one must keep in mind that the SGWs are distinct from the PGWs and that there is a Visited SGW which allows to connect the Visited MME to the Home PGW through the GRX network. The Visited SGW as well as the Visited MME must be known and addressable by the Home PGW, for example, to change the Qos on demand of the external services used by the subscribers. This is why the GTP v2 protocol used in LTE or recent 3G is much more complicated than GTP v0, v1 used in 2.5G or older 3G.

Functions of the MME (Mobility management entity), an evolution of a SGSN 3G, include:

-   NAS ;

- Inter CN node signaling for mobility
- UE Reachability and Tracking Area list management;
- PDN GW(mobile User Equipment) and Serving GW selection;
- MME selection for handovers with MME change;
- SGSN selection for handovers to 2G or 3G 3GPP;
- Roaming (S6a towards home HSS);
- Authentication;
- Bearer management functions including dedicated bearer establishment.
- Lawful Interception of signaling traffic.

SGW (Servicing Gateway) includes

- Mobility anchoring for inter-3GPP mobility
- ECM-IDLE mode downlink packet buffering and initiation of network triggered service request procedure;
- Lawful Interception;
- Packet routing and forwarding;
- Transport level packet marking in the uplink and the downlink based on the QCI (Quality Class Identifier) of the associated EPS bearer;
- Accounting on user and QCI granularity for inter-operator charging;
- UL and DL charging per UE, PDN, and QCI (e.g. for roaming with home routed traffic).
- Interfacing OFCS according to charging principles.

PGW (Packet Data Network Gateway) includes

- Per-user based packet filtering (by e.g. deep packet inspection);
- Lawful Interception;
- UE IP address allocation;
- Transport level packet marking in the uplink and downlink;
- UL and DL service level charging ;
- Interfacing OFCS through according to charging principles.
- UL and DL service level gating control as defined in TS 23.203
- UL and DL service level rate enforcement as defined in TS 23.203
- UL and DL rate enforcement based on APN-AMBR;
- DL rate enforcement based on the accumulated MBRs of the aggregate of SDFs with the same GBR QCI;
- DHCPv4 (server and client) and DHCPv6 (client and server) functions;

- The network does not support PPP bearer type in this version of the specification;
- packet screening.

The PG-W includes a PCEF able of protocol analysis with Deep Packet Inspection (DPI). It interrogates the PCRF to obtain individual credit and QoS information in order to accept the data connections and allocate a quality of service based on the traffic needs (for VoIP a small delay is necessary which is not for email).

PCRF (Policy and Charging Rules Function) includes
PCRF is the policy and charging control element (pre-paid and postpaid)
- H-PCRF that resides within the Home-PLMN,
- V-PCRF that resides within the Visited-PLMN.

PGW's associated AAA Server
The PDN Gateway may interact with an AAA server over the SGi interface. This AAA Server is used for other network such as Wifi or Wimax.

PROTOCOLS used in the LTE network
The EPC uses the standard **GTP** (same as in 3G) between the MME and the PDN gateway. The signaling plane uses **DIAMETER protocol replacing the MAP protocol:**

- **S6a** interface between the MME and the HSS (3GPP TS-29272) used for mobility management,
- **X2** interface between eNodeB (e-utran Node B) ,
- **Gx** interface between PDN-GW (Packet Data Network Gateway) and PCRF (Policy and Charging Rules Function),
- **Cx** interface between CSCF and HSS (Home Subscriber Serve) (3GPP TS-29229)
- **X2** interface used by a eNB to communicate to other eNB with an IP interface with SCTP as transport.
- **S1** the eNode B and MME communicate using the S1-MME protocol, while the eNode B communicates with the SGW using the S1-U protocol. The transport protocol is stream control transmission protocol (SCTP).

3.7.2 Interface Protocols of a PDN Gateway

A full PDN Gateway development is a major undertaking because of the number of protocols used as shown by Table 3.4.

Table 3.4 Standard interfaces of a PDN Gateway

Name of interface	Interfacing with	Protocol	Specification
S5/S8 (GTP implementation)	SGW	GTP v2/UDP/	3GPP TS 29.274
S5/S8 (PMIP implementation)	SGW	PMIP/IP	3GPP TS 29.275
Gx	PCRF (PCEF in PGW)	Diameter/SCTP/IP	3GPP TS 29.212
Gy	Online Charging	Diameter/SCTP/IP	3GPP TS 32.240
Gz (Ga implementation)	Offline Charging	GTP'/UDP or TCP)	3GPP TS 32.299
Gz (Rf implementation)	Offline Charging	Diameter	3GPP TS 32.295
SGi (equivalent Gi in 3G)	External Networks	TCP(browsing)or UDP	
S2a	TWLAN (trusted WLAN)	PMIP/IP or MIPv4/UDP/IP	3GPP TS 29.275
S2c	ePDG (non trusted WLAN)	DSMIPv6, IKEv2	3GPP TS 24.303

For a GTP Hub, only the protocols in grey would need to be implemented. A local PCRF will be used. The charging will use the RADIUS server integrated with the PGW with a total traffic credit for a given visitor's MSISDN (the RADIUS server is not shown in the interfaces).

3.7.3 General LTE Call Flow

To understand how LTE is working, here is the full, not detailed, call flow for the LTE attach and default Bearer Setup. When the user is in the reach of a LTE network, his UE will "attach" which means than an UPDATE

LOCATION with his IMSI will be sent to his HSS (the equivalent of the HLR in 3G), to get the APN and some QoS parameters, very much like the "attach" to a 3G UTRAN network. But in the LTE, there is also the creation of the permanent connection with a default QoS. The procedure is then called "Attach" (the updating of the location information in the visited MME) and "Default Bearer Setup" creation (using the GTPv2 Create Session Request which is the equivalent of Create PDP Context Request in 3G).

But when the UE effectively initiates the connection to a network service, it will want to update the QoS in the "default bearer context" which is not appropriate. It will trigger in the MME the sending of an Update Bearer Request to the SGW to update the" bearer context". In 3G GTPv1, the equivalent is Update PDP Context which can change the QoS.

3.7.3.1   Detailed Call Flow of the Session Creation

Detailing further the call flow of the Create Session Request allows us to understand the separation of functions between the MME and the SGW (which is not a transparent relay between the MME and the PGW for the GTPv2 signaling messages, but which is a transparent router for the GTP-U traffic data). The MME does not handle the GTP-U user data traffic, only the SGW does, while the PGW handles both.

One can see that in the Create Session Request the MME is providing the SGW with the Tunnel End Point Identifier of the eNodeB for the GTP-U data traffic. The SGW replaces the IP address of the eNodeB by its own. In the Create Session Response you can see that the SGW also replaces the Tunnel IP addresses for the Control Plane and the User Plane. The SGW is not a transparent router for the GTP-C traffic.

To implement a GTP v2 Hub, only the PGW function is necessary. To implement a medium size PGW in a network, it is economical to also integrate the SGW as a distinct function.

3.7.3.2   Traces of the Session Creation with the Setting of the Initial Default "Bearer Context"

This trace illustrates the case of Figure 3.6, an LTE outbound subscriber establishing a data session to an external service through his home PGW (assuming there is no LBO). The Create Session Request is sent from the MME to the home PGW through the visited network SGW which relays the Create Session Request after adding the User Plane (GTP-U) information using the GRX. This trace is taken at the S11 interface of the visited MME. It does not yet (will be added by the visited SGW) contain the User Plane address and Tunnel End Point Identifier for GTP-U.

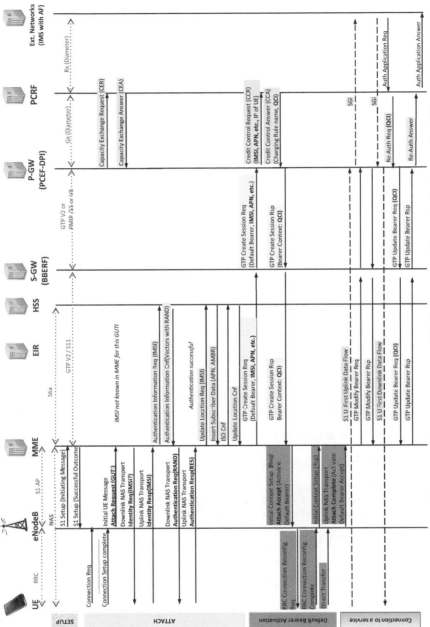

Figure 3.6 LTE Attach, Initial default Bearer Setup and subsequent QOS change requested by an Application Function (AF).

Header IP received _____

    Src IP '192.168.2.2'

    Dst IP '192.168.0.4'

    - - - - Super Detailed GTP, Gi, SIP, DIAMETER, RADIUS Analyser (C)HALYS - RS=1 ENCAPS=1 - -

        (48)GTP version 2(1=GSM,2= LTE), T(presence TEID)=1

        Message Type:(32):Create Session Request (MME(S11+S5/S8) or SGSN(S4+S5/S8) or TWAN(S2a) or ePDG(S2b)-> PGW)

        Length_Payload 220

        TEID 0xA8A3A2A1

        SequenceNumber 1234567

        Spare 00

         (1):IMSI(International Mobile Subscriber Identity)

         L = 8

         CR flag: 00 Instance: 00

         Data: 208017001772673

         (76):MSISDN

         L = 6

         CR flag: 00 Instance: 00

         Data: 37493600800

         (75):Mobile Equipment Identity(MEI)

         L = 8

         CR flag: 00 Instance: 00

         Data: 123456123456789

         (86):User Location Information(ULI)

         L = 13

         CR flag: 00 Instance: 00

         TAI(Tracking Area Identity) present(5 octets)

          MCC 293 MNC 5 Tacking Area code 2

         ECGI present(7 octets)

          MCC 293 MNC 5 Spare 8 E-Utran Cell Indentifier(ECI 3 octets) 123456

         (83):Serving Network(SN)

          L = 3

         CR flag: 00 Instance: 00

         Data: MCC = 283 MNC = 5

         (82):Radio Access Technology(RAT)

          L = 1

         CR flag: 00 Instance: 00

         Data: (6):EUTRAN

         (87):Fully qualified End Point Identifier(F-TEID)

         L = 9

         CR flag: 00 Instance: 00

         messtype 32

         IP type: (2):IPv4

         Interface type: (10):S11 MME GTP-C interface (Local or BackBone GRX IP of Visited MME)

        TEID/GRE Key: 0

        IPv4: 1.2.3.4

         (87):Fully qualified End Point Identifier(F-TEID)

         L = 9

         CR flag: 00 Instance: 00

         messtype 32

         IP type: (2):IPv4

Interface type: (7):S5/S8 PGW GTP-C interface (BackBone GRX IP of Visited SGW)

TEID/GRE Key: 0
IPv4: 2.3.4.5
(71):APN(Access Point Name)
L = 12
CR flag: 00 Instance: 00
Data: xxx.yyy.zzz
(128):Selection Mode of APN
L = 1
CR flag: 00 Instance: 00
Data: (2):Network provided APN,subscription not verified
(99):PDN Type
L = 1
CR flag: 00 Instance: 00
PDN type: (1):IPv4
(79):PDN Address Allocation(IP allocated to UE)
L = 5
CR flag: 00 Instance: 00
Data: 123.124.125.126
PDN Type (1):IPv4
(127):APN Restriction
L = 1
CR flag: 00 Instance: 00
Data: 2
(72):Aggregate Maximum Bit Rate(AMBR)
L = 8
CR flag: 00 Instance: 00
Uplink: 70000 bits/sec
Downlink: 200000 bits/sec
(78):Protocol Configuration Options(PCO)
L = 23
CR flag: 00 Instance: 00
Data: 80802110010000108106000000000830600000000000A00
(93):Bearer Context
L = 31
CR flag: 00 Instance: 00
Data: 4900010005500016005C06000000000000000000000000000000000000000000
  (73):EPS Bearer ID (EBI)
  L = 1
  CR flag: 00 Instance: 00
  Data: 5
  (80):Bearer level Quality of Service
  L = 22
  CR flag: 00 Instance: 00
  Pre-emption Vulnerability: (0):Enabled
  Priority Level: 7
  Pre-emption Capability: (1):Disabled
  Label (QCI): (6):Non-GBR,priority = 7, 100 ms: Interactive gaming
  Maximum Bit Rate for Uplink: 0.000000
  Maximum Bit Rate for Downlink: 0.000000
  Guaranteed Bit Rate for Uplink: 0.000000
  Guaranteed Bit Rate for Downlink: 0.000000

```
(3):Recovery Restart Counter
 L = 1
 CR flag: 00 Instance: 00
 Data: 1
(95):Charging characteristics
 L = 4
 CR flag: 00 Instance: 00
 Data: 0102621F
```

In the Create Session Response, the User Plane addressing information (S1-U SGW GTP-U interface) is set by the PGW in the Bearer Context parameter. So that the MME can be ready to start the Uplink Data Flow in Figure 3.6 as soon as the MME has received both this Create PDP Session Response and the « Attach Complete » sent by the eNodeB.

Figure 3.7 shows two important things:

- the GTP-C changes through the SGW (if any , case of large networks, several PGW),
- the fact that the Create Session Request sent by the MME *does not include any User Plane addressing information* for the eNodeB. This an important difference with SGSNs in 3G where the Create PDP Context sent by the *SGSN included both its Control Plane and its User Plane addressing* information(to receive down link flows to the MS).

```
_____ Header IP sent _____
 Src IP '192.168.0.4'
 Dst IP '192.168.2.2'
 - - - - Super Detailed GTP, Gi, SIP, DIAMETER, RADIUS Analyser (C)HALYS - RS=0
ENCAPS=1 - -
       (48)GTP version 2(1=GSM,2= LTE), T(presence TEID)=1
       Message Type:(33):Create Session Response (MME or SGSN or TWAN or ePDG<-
PGW)
       Length_Payload 91
       TEID 0x00000000
       SequenceNumber 1234567
       Spare 00
         (2):Cause
          L = 2
          CR flag: 00 Instance: 00
          Data: (16):Request Accepted
         (93):Bearer Context
          L = 37
          CR flag: 00 Instance: 00
          Data:
0200020010004900010005570009008500000008C0A800045700090081000000008C0A80004
         (2):Cause
          L = 2
          CR flag: 00 Instance: 00
```

Data: (16):Request Accepted
(73):EPS Bearer ID (EBI)
  L = 1
  CR flag: 00 Instance: 00
  Data: 5
(87):Fully qualified End Point Identifier(F-TEID)
  L = 9
  CR flag: 00 Instance: 00
  IP type: (2):IPv4
  Interface type: (5):S5/S8 PGW GTP-U interface (BackBone GRX IP of Home GGSN-PGW)
    TEID/GRE Key: 00000008
    IPv4: 192.168.0.4
  (87):Fully qualified End Point Identifier(F-TEID)
  L = 9
  CR flag: 00 Instance: 00
  IP type: (2):IPv4
  Interface type: (1):S1-U SGW GTP-U interface (Local or BackBone GRX IP of Home SGW)
    TEID/GRE Key: 00000008
    IPv4: 192.168.0.4
  (3):Recovery Restart Counter
  L = 1
  CR flag: 00 Instance: 00
  Data: 1
(87):Fully qualified End Point Identifier(F-TEID)
  L = 9
  CR flag: 00 Instance: 00
  messtype 33
  IP type: (2):IPv4
  Interface type: (11):S11 or S4 SGW GTP-C interface (Local or BackBone GRX IP of Home SGW)
    TEID/GRE Key: 1
    IPv4: 192.168.0.4
  (87):Fully qualified End Point Identifier(F-TEID)
  L = 9
  CR flag: 00 Instance: 00
  messtype 33
  IP type: (2):IPv4
  Interface type: (170):S5/S8 PGW GTP-C interface (BackBone GRX IP of Home GGSN-PGW)
    TEID/GRE Key: 8
    IPv4: 192.168.0.4
  (79):PDN Address Allocation(IP allocated to UE)
  L = 5
  CR flag: 00 Instance: 00
  Data: 123.124.125.126
  PDN Type (1):IPv4
(131) Change Reporting Action
  L = 4
  CR flag: 00 Instance: 00
  Data: (1):start rporing CGI/SAI
        (2): start reporting RAI

(3): start reporting TAI
(4): start reporting ECGI
(165) H(e)NB Information Reporting
L = 1
CR flag: 00 Instance: 00
Data: (1): FTI on: start reporting when the UE moves from an eNodeB to another

The role of the last two parameters in italic is explained in the section 13.4.3.2 on Mobile Advertisement services in LTE networks.

3.7.3.3   Creation of the Default Dowlink Bearer QoS and Start of the Downlink Data Flow

To initiate the First Downlink Data Flow (Figure 3.6) the MME *must send* a GTP Modify Bearer Request to the PGW which contains a default QoS and also the User Plane addressing information of the eNodeB, that is « GTP-U S1-U eNodeB TEID and IP address ». When received, the PGW unblocks the Downlink Data for the concerned UE as it knows now how to send the GTP-U messages to the eNodeB.

3.7.3.4   Dialog between the PCEF-DPI and the PCRF (Gx/DIAMETER)

When the PGW-GGSN receives the Create Session Request, it needs the QoS-Class-Identifier (QCI) to fill the AVP Bearer Request in the Create Session Response. According to the value (1–9), the eUTRAN will set the radio QoS allocated to the concerned subscriber.

With the Credit-Control-Request (CCR), the PCEF-DPI asks this QCI to the PCRF. It provides the following AVPs
-   IMSI
-   MSISDN
-   APN

and
-   Framed-IP-Address (the IP which has then been assigned by the DHCP at this stage)
-   a Session-Id assigned by the PCEF-DPI.

The PCRF responds with the AVP Charging-Rule-Name which is a template profile for the QCI, the DL and UL minimal bandwidth, etc., but does not include the QCI which is then implicitly based on the Charging-Rule-Name preconfigured between the PCEF and PCRF. This provides much more information than just the QCI value which could be returned in the Credit-Control-Answer (CCA) by the PCEF-DPI in the AVP
  Qos-Information/QCI.

To clarify the logic of the overall call flow 3.5, we have assumed that the PCRF returns explicitly the QCI parameter (the value received from the

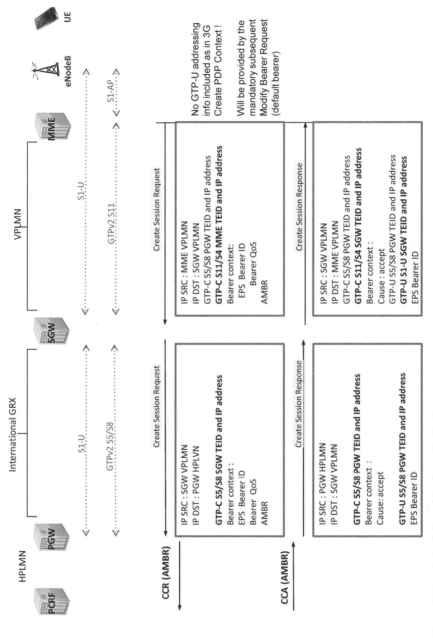

Figure 3.7 Detailed call flow of the Create Session Request

PCRF is included in the AVP Bearer Request). The QCI is then either provided explicitly in the AVP Qos-Information/QCI or implicitly because it is defined in the AVP Charging-Rule-Name.

3.8 APN Correction Service in a GTP Hub Using CAMEL

A much more complicated alternative, where use of a GTP Hub avoids the change of APN, is the use of "IN service" if all the visitors have CAMEL SGSN subscriptions. In this case, the SGSN will send an INITIAL DP GPRS request to the proxy SCP *before* the Create PDP Context.

The "IN service" in the HPLMN may correct the APN and replace it, for example, by EUInternet which is *supposed to be in the list of authorized APNs* sent by its HLR to the SGSN. This way, the SGSN will control and accept to make the Create PDP Context to the normal GGSN (no need for a GTP Hub).

This requires, however, that the HPLMN (it provides the IN service from one of his SCPs) and the VPLMN agree to the LBO of the visitor, which is doubtful! If this is not the case, and if the VLMN still prefers to use its ordinary GGSN, it may use a GLR which will be able to add a specific CAMEL trigger in the SGSN profile. The traces for APN Correction by CAMEL performed by the IN service in the HPLMN (commercially doubtful) or in a VPLMN GLR are below.

3.8.1 CAMEL Sent by the VPLMN SGSN

```
- - - - Super Detailed SS7 Analyser (C)HALYS - - - - - - -
TCAP      CONTINUE
TC_INVOKE(Request operation(Last))
TCPPN_LAST_CPT(2)
L = 001
Data: 1
TCPPN_COMPONENT(1)
L = 092
Data: Component type: Invoke(A1)
Invoke ID(2)
L = 001
Data: 0
CAP-INITIAL-DP-GPRS(78)
        CAPTag_Service Key(128)
        L = 001
        Data: 56 /* Service Key in the SCF for the « APN correction service »
        CAPTag_Event_Type_GPRS(129)
    L = 001
        Data: (11): pdp-context Establishment, /* "trigger" set by the HLR */
        CAPTag_MSISDN(130)
        L = 007
```

Data: Ext = No extension
Ton = International
Npi = ISDN
Address = 35465XXXXXX
CAPTag_imsi(131)
L = 008
Data: Address = 27404XXXXXXXXXX
CAPTag_Time_and_TimeZone(132)
L = 008
Data: 2012.11.24 05:34:30 04
CAPTag_Access Point Name(136) /* telephone configured avec cet APN(NI)
L = 009
Data: **viking.is /* standard APN Internet to be changed to EUinternet**
CAPTag_locationinformationGPRS)(172)
L = 018
Data: (Hex) 800702F801B5A4460F8307913306090011F0
Cell ID (in the CS network)(128
L = 007
Data: **MCC = 208 MNC = 10 LAC = 46500, Cell_ID = 17935**
SGSN_number(131)
L = 007
Data: Ext = No extension
Ton = International
Npi = ISDN
Address = 33609000110 /* SGSN SFR visited by the mobile which is "CAMELized"
CAPTag_PDP Initiation type(141)
L = 001
Data: (0):MS Initiated
CAPTag_IMEI(145)
L = 008
Data: Address = **3593200218568908 /* the IMEI is available */**

3.8.2   Response from the IN APN Correction Service

2
‾
- - - - Super Detailed SS7 Analyser (C)HALYS - - - - - - -
TCAP END
TC_INVOKE(Request operation(Last))
TCPPN_LAST_CPT(2)
L = 001
Data: 1
TCPPN_COMPONENT(1)
L = 092
Data: Component type: Invoke(A1)
Invoke ID(2)
L = 001
Data: 0
CAP-CONNECT-GPRS(74)
         CAPTag_Access Point Name(136)
         L = 009
         Data: NI= *euinternet* /* **APN Internet (Network Indicator)changed to EUinternet by the**
IN service

After receiving this,

- the SGSN checks that *euinternet* is in the list of APN received from the HLR in the GPRS profile received from the GLR,
- makes a local DNS request and finds the local GGSN or GTP Hub address,
- establishes a PDP context with the local GGSN and the LBO service is given.

Another strategy is when the HPLMN cooperates for the LBO.

Above, the APN sent by the IN service does not change the OI (Operator Identifier) part of the APN, so that the SGSN will use the following address for the DNS resolution:

euinternet.mnc004.mcc274.gprs

The LBO will be performed if this APN has been included in the list [3.1]. If this is not the case, knowing that the VPLMN has MCC-MNC=285-05, the HPLMN may compel the service by forcing the use of the VPLMN GGSN/PGW by changing also the OI field.

CAP-CONNECT-GPRS(74)
     CAPTag_Access Point Name(136)
     L = 009
     Data: NI= *euinternet* **OI=mnc005.mcc283.gprs /* APN Internet changed to EUinternet and use of the local GGSN/PGW by the IN service**

All this is to illustrate the flexibility of CAMEL for data services than to illustrate a probable business case. The HPLMNs will not be happy with the LBO service unless they have many visitors in their own network and can compensate with additional data service revenue.

3.9 MMS Anti-Spam Service [3.2]

3.9.1 Real MMS Anti-Spam: Content Retrieval Mandatory

We want to implement a *content-based* MMS anti-spam. The problem is much more complicated than SMS anti-spam as even the filtering based on the "calling party" would need to receive the MMS and open it as the *calling party is not in general in the MMS notification by SMS* where the origin number is alphanumeric ("MMSC") or the same for all originators with most MMSC vendors. Claims by some SMS anti-spam vendors that they can also control the origin of the MMS based on the SMS containing the MMS notification are then incorrect.

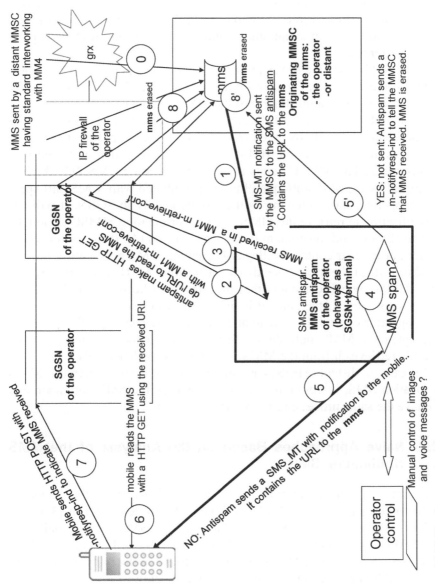

Figure 3.8 MMS anti spam system.

The GTP Hub technology which includes an "SGSN side" is applicable to implement a real MMS anti-spam system such as Figure 3.4, which includes a mobile terminal + SGSN simulator. It behaves as the destination handset to retrieve the MMS *before it reaches the handset*, extract the real "calling party" (field From, example +3360812346/TYPE=PLMN), and the content so that even an automatic anti-spam algorithm of the text parts may be performed.

Refer to [3.3] for the MMS protocols (MM1 for the mobiles and MM4 for MMS interworking). Chapter 7 explains the MMS Hub architecture for interworking through a third party of the MM4 protocol.

Figure 3.8 explains the MMS anti-spam system. The MMS containing eventual spam is usually received with the SMTP-based protocol MM4 (0) and arrives in the destination MMSC of the operator which will send it to his subscriber. The MMSC sends a MMS Notification SMS (1) which is intercepted by the SMS anti-spam function combined with MMS anti-spam that the operator has. Behaving as the mobile and as an SGSN, it will use the notification to retrieve the MMS from the originating MMSC (2 and 3).

The "From" and the content (text) are analyzed (4). If the calling party is barred or the text content is illegal(4) then the MMS notification SMS is not sent, instead the system sends (5') a positive MM1 m-notify-resp-ind to the MMSC which considers that the MMS has been sent and received and does not retry. If the content includes images or sounds, it can be sent to the operator control for a manual check. On the contrary, if the content is accepted the MMS notification SMS is sent to the mobile (5) which normally retrieves it from the MMSC (6) and acknowledges (7).

In case a distant MMSC is attempting to send directly a MMS Notification SMS, it would also be intercepted by the SMS anti-spam and will have the same processing for MMS anti-spam.

3.9.2 Naive Approaches Based on the Analysis of the MMS Notification SMS

Some have described systems that would at least filter the "Sender" number assuming that the MMS notification MMS described in [3.3] and [3.4] has the sender in the Origin Address of the SMS. This is naive, as most MMSC use a single address (MMSC, MMSCviva, 1234) for all these SMS. The sender +3360123456 can only be found in the body of the MMS. So retrieval before filtering is mandatory. Trace below:

......
X-Mms-Message-Type: MM1-m-retrieve-conf
X-Mms-Transaction-ID: "20110418.18.2C2BFBCA-p33617554874-025727"

X-Mms-Message-ID: "20110418.18.2C2BFBCA@192.168.0.4-025727"
To: +354389123456/TYPE=PLMN
From: +3360123456/TYPE=PLMN
X-Mms-Originator-System: system-user@mms.sfr.fr
Sender: +3360123456/TYPE=PLMN@mms.sfr.fr
X-Mms-Priority: High
X-Mms-Ack-Request: Yes
X-Mms-Delivery-Report: Yes
X-Mms-Read-Reply: No
Message-ID: <Msg-20111109103341-82366-0976186416>
Content-Type: multipart/related; start="<4B9FC4C2.smil>"; type="application/smil";
boundary="---mime-boundary-7E23F647.5E4F1808"

-----mime-boundary-7E23F647.5E4F1808
Content-Type: application/smil; name="4B9FC4C2.smil"; charset=utf-8
Content-ID: <4B9FC4C2.smil>
Content-location: 4B9FC4C2.smil

...

References and Further Reading

[3.1] A. Henry-Labordère, S.Cruaux, Gilles Duporche "Evasion   IP   locale
dynamique pour les mobiles en itinérance", Patent 13 51 461
[3.2] A.Henry-Labordère, G..Duporche, W.Manaï, B.Mathian,"Système MMS
anti-spam", Patent 12 54 308
[3.3] ETSI TS 123 140 V6.16.0(2009-04), "Digital cellular telecommunications
system (Phase 2+); Universal Mobile Telecommunications System (UMTS);
Multimedia Messaging Service (MMS); Functional description; Stage 2"
(3GPP TS 23.140 version 6.16.0 Release 6)
[3.4] openmobilealliance.org, WAP-209-MMSEncapsulation-2020105a, "Wireless
Application Protocol, MMS encapsulation protocol"
[3.5] openmobilealliance.org, OMA-WAP-TS-ProvSC-V1_1-20090728-A, "Open
Mobile Alliance, Provisioning Smartcard".
[3.6] H.Holma and Antti Toskala editors, "LTE for UMTS, Evolution to LTE-
Advanced", Wiley ed., 2011.
[3.7] ETSI TS 129 060 V11.7.0 (2013-6), "Digital cellular telecommunications
system"   (Phase   2+);Universal   Mobile   Telecommunications   System
(UMTS);General Packet Radio Service (GPRS);GPRS Tunneling Protocol
(GTP) across the Gn and Gp interface (3GPP TS 29.060 version 11.7.0
Release 11).
[3.8] ETSI TS 129 274 V11.7.0 (2013-7), "Universal Mobile Telecommunications
System"   (UMTS);LTE;3GPP   Evolved   Packet   System   (EPS);Evolved
General Packet Radio Service (GPRS) Tunneling Protocol for Control plane
(GTPv2-C);Stage 3 (3GPP TS 29 274 version 11.7.0 Release 11)

4

Radius Hub (national and international IP roaming) and secured Radius-based data communications

4.1 The Radius Protocol: Tutorial and Applications

The RADIUS protocol possibilities are sometimes not well known. This is a UDP/IP protocol (not connected) with not many different messages. The parameters are called "AVPs" (Attribute Value Pairs). There are three main sets:

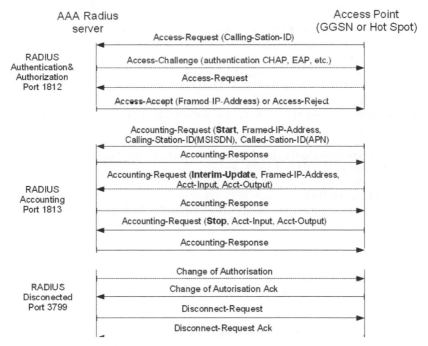

Figure 4.1 Radius message flow.

Authentication and Authorization [4.3]. This is not generally configured between a GGSN and the RADIUS AAA server, because the subscriber has already be authenticated by the SS7 UPDATE LOCATION GPRS.

Accounting [4.4]. This is always used irrespective of whether the NAS is a hotspot or a GGSN. In the GGSN case, it is necessary for certain services so that the AAA server has the pair "Framed-IP-Address", the IP allocated to the MS and the Calling Party Id (the MSISDN).

Change of Authorization [4.6]. This allows to disconnect and release an IP connection or to change authorizations.

4.1.1 Analysis of an Access-Request

If a WiFi hotspot access point is used as in Chapter 8, an authentication is absolutely necessary. Many methods may be used, in particular, the user name + password which refer to the CHAP method. The example below corresponds to the EAP-SIM method which works with a SIM card in the terminal

```
- - - - Super Detailed GTP, SIP, DIAMETER, RADIUS Analyser (C)HALYS - - - - - - -
message length = 388
_____ Header RADIUS (port 1812): RECEIVED _____
RADIUS Command(1)Access-Request
Packet identifier 14
headerRADIUS(with Authenticator)+payload = 388
Authenticator 431E7B45C8C9B7D8552A578EB3F88CA7 (MD5 calculated using shared
secret: totoalaplage)
 (50):Acct-Multi-Session-ID(Key unique for authentication session: MAC Hotspot|Mac
Terminal|sequence)
length AVP = 61
value = 2C-41-38-F1-58-D0-04-1E-64-79-AD-47-50-2B-97-70-00-02-AA-E4
(44):Acct-Session-Id(Call ID of the SIP session)
length AVP = 19
value = 6435141c-0000182c
(5):NAS-Port(physical connection on NAS,NOT the TCP or UDP 'port number')
length AVP = 6
value = 6096
(61):NAS-Port-Type
length AVP = 6
(19):Wireless IEEE 802.11
(32):NAS-Identifier(ID of the Hotspot Wifi)
length AVP = 12
value = CN17DWZ3W7
(4):NAS-IP-Address(GGSN Radius IP, Hotspot WiFi IP)
length AVP = 6
value = 192.168.1.17
(12):Framed-MTU(max packet size)
```

length AVP = 6
value = 1496
(1):User_Name(e.g.set in the softphone)
length AVP = 53
MSISDN = 1623020000660004
value = 1623020000660004@wlan.mnc002.mcc623.3gppnetwork.org
(31):Calling-Station-Id:MSISDN of handset(GPRS or EAP-SIM) or MAC address of Terminal(Hotspot WiFi)
length AVP = 19
value = 04-1E-64-79-AD-47
(30):Called-station-Id(APN selected by handset(GPRS) or MAC adress of Hotspot(WiFi)
length AVP = 19
value = 2C-41-38-F1-58-D0
(6):Service-Type
length AVP = 6
(2):Framed-User
(79):EAP-Message(NAS->AAA)(see RFC 4186)
length AVP = 58
value of AVP after concatenation(L= 56):
028900380131363233303230303030303636363030303440776C616E2E6D6E633030322E6D6
3633632332E336770706E6574776F726B2E6F7267
Code: (2):Response
Identifier: 137
Length: 56 octets
Type: (1):Identity(IMSI Based)
Network Access Identifier= 1623020000660004@wlan.mnc002.mcc623.3gppnetwork.org (L=51):
first authentication method requested by mobile = (49):EAP-SIM
username(IMSI) = +623020000660004
(80):Message-Authenticator(see RFC 3579) MANDATORY for EAP in NAS->AAA and AAA->NAS
length AVP = 18
value = 7AEA517F1D6D7A9E64790BC4CE9B1AD1 (HMAC-MD5 calculated with shared secret: totoalaplage)

Note that all RADIUS headers have a mandatory Authenticator in the header (431E7B45C8C9B7D in the example) which is a signature computed by hash function MD5 over the entire message with a shared secret between the access point and the AAA server. There is also an optional Message-Authenticator to further prevent a spoofing attack against any of the two.

We can see that the Calling-station-ID is the MAC of the terminal and the Called-Station-ID is the MAC of the hotspot. Since it is EAP-SIM, there is a special AVP "EAP-Message" in the EAP-SIM protocol that includes the identity of the terminal:

1623020000660004@wlan.mnc002.mcc623.3gppnetwork.org

This includes the indication 1 for EAP-SIM, the IMSI 623020000660004 and the "realm" identifying the HPLMN (computed by the terminal from the IMSI).

4.1.2 Analysis of an Access-Accept with Allocation of the IP Address

The AAA server will Accept or return Access-Failure, but a feature of the RADIUS protocol is that it can assign the IP address (172.17.10.229 in the example) of the terminal (the GGSN or the Hot Spot will not perform a DHCP request). Some MNOs use this RADIUS IP allocation instead of a DNS query from the GGSN. The RADIUS has its own pool of addresses or performs the access to a DHCP server. To work, the GGSN or the Hot Spot must support this IP allocation by RADIUS (rare for standard Hot spots).

```
_SENT____ Header IP SENT _____
- - - -Super Detailed GTP, SIP, DIAMETER, RADIUS Analyser (C)HALYS - - message length = 193
____ Header RADIUS (port 1812): SENT ____
RADIUS Command(2)Access-Accept
Packet identifier 5E
headerRADIUS(with Authenticator)+payload = 193
Authenticator   02C4BB641F14491EC09422CB8E243535   (MD5   calculated   using   shared   secret:
totoalaplage)
 (79):EAP-Message(NAS->AAA)(see RFC 4186)
length AVP = 6
value of AVP after concatenation(L= 4):
038C0004
Code: (3):Success
Identifier: 140
Length: 4 octets
(27):Session-Timeout(max duration of session(seconds))
length AVP = 6
value = 25
(28):Idle-Timeout(max idle time of session(seconds))
length AVP = 6
value = 60
(42):Acct-Input-Octets(NAS->Terminal CUMULATED)
length AVP = 6
value = 10
(43):Acct-Output-Octets(Terminal->NAS CUMULATED)
length AVP = 6
value = 10
(85):Acct-Interim-Interval(Interval  between  Accounting-Request  'Interim-Update'  NAS->AAA)  in
seconds)
length AVP = 6
value = 15
(80):Message-Authenticator(see RFC 3579) MANDATORY for EAP in NAS->AAA and AAA->NAS
length AVP = 18
value  =  9F3A44D1F27E54A784357F3CF6F82779  (HMAC-MD5  calculated  with  shared  secret:
totoalaplage)
```

(26):Vendor-Specific(3GPP TS 29.061)
Vendor_ID = 311 ((17):MS-MPPE-RECV-KEY(to cypher the AP->station)
salt   =   C5A2   MS-MPPE-Key   =
1B240E35B8886E61CB3A5E10AF2A10B25B98F64E18D32AA7F75F47321E330164C6BC465C71F
857ECDEDE0DDE820ACBC7
(26):Vendor-Specific(3GPP TS 29.061)
Vendor_ID = 311 ((16):MS-MPPE-SEND-KEY(to cypher the station->AP))
salt   =   C5A3   MS-MPPE-Key   =
C22E8AE890FCED1F0EE22B63296ADEBF8A3C2AE40CD87AA1A4E09378570E34342C140E07E
FC398863764A6BDD58D5F58
(8):Framed-IP-Address(IP of handset assigned by DHCP);presence means Terminal is IP
ACTIVATED!
length AVP = 6
value = **172.17.10.229**

Notice two "Vendor-Specific" Microsoft AVP MPPE-RECV-KEY and
MPPE-SEND-KEY which are standard as this is a Hot Spot access
example. They are used for a common WiFi ciphering between the
terminal and the Hot Spot, as used in Chapter 8.

4.1.3   Analysis of an Accounting Request (GGSN Example)

The trace below gives details of the parameters in an Accounting-Request.
This is always used by Access Points (WiFi hot spots or GGSN). We see
that the IP-Framed-Address, the Calling-Station-ID, the APN and the IMSI
are passed, in general, by a GGSN or PGW to the AAA server

_____ Header RADIUS (port 1813): RECEIVED _____
RADIUS Command(4)Accounting-Request
Packet identifier 1F
headerRADIUS(with Authenticator)+payload = 392
Authenticator 454BD6C96BAC546E85637EF41FEBBD87 (MD5 calculated using shared secret:
imc#halys#vodaphone)

(50):Acct-Multi-Session-ID(Key unique for authentication session: MAC Hotspot|Mac
Terminal|sequence)
length AVP = 10
value = ee47102d
(51):Acct-Link-Count
length AVP = 6
value = 00000001
(55):Event-Timestamp(NAS->AAA)
length AVP = 6
value = 50F00DCC
(31):Calling-Station-Id:MSISDN of handset(GPRS or EAP-SIM) or MAC address of Terminal(Hotspot
WiFi)
length AVP = 12
value = 3548698264
(8):Framed-IP-Address(IP of handset assigned by DHCP);presence means Terminal is IP
ACTIVATED!
length AVP = 6

value = **172.17.10.229**
(44):Acct-Session-Id(Call ID of the SIP session)
length AVP = 18
value = c2902093ee47102d
(4):NAS-IP-Address(GGSN Radius IP, Hotspot WiFi IP)
length AVP = 6
value = 10.153.252.241
(7):Framed-Protocol
length AVP = 6
(7):GPRS-PDP-Context
(45):Acct-Authentic
length AVP = 6
(2):Local
(30):Called-station-Id(APN selected by handset(GPRS) or MAC adress of Hotspot(WiFi)
length AVP = 12
value = mms.imc.is /* APN */
(32):NAS-Identifier(ID of the Hotspot Wifi)
length AVP = 7
value = G01.c
(41):Acct-Delay-Time(delay added in the NAS for Accounting-Request NAS->AAA) in seconds
length AVP = 6
value = 0
(61):NAS-Port-Type
length AVP = 6
(5):Virtual
(6):Service-Type
length AVP = 6
(2):Framed-User
(40):Acct-Status-Type(Start,Stop,Interim-Update,etc..)
length AVP = 6
(1):Start

The accounting-Request setup in the GGSN may include many useful so-called Vendor-Specific AVPs such as:

(26):Vendor-Specific(3GPP TS 29.061)
Vendor_ID = 10415 ((1):IMSI of handset)
value = 274040199000094
(26):Vendor-Specific(3GPP TS 29.061)
Vendor_ID = 10415 ((6):SGSN-IP Address(visited by handset))
value = 157.157.144.65
(26):Vendor-Specific(3GPP TS 29.061)
Vendor_ID = 10415 ((7):GGSN-IP Address(used by handset))
value = 194.144.32.147
(26):Vendor-Specific(3GPP TS 29.061)
Vendor_ID = 10415 ((20):IMEISV(IMEI+SW Version))
value = 3562990476509501

The AAA server is then interrogated by equipments which need more information. For example, an MMSC receives an MMS request from a handset. It knows only the Src IP address which is the "Framed-IP-Address". With this, it interrogates the RADIUS AAA server and can obtain the necessary MSISN (Calling-Station-ID) and also, if needed for reformatting, the IMEI. All these parameters have been obtained by the GGSN in the Create PDP Context.

4.1.4 Use of the Disconnect-Request

In a GSM network using a PCRF and PCEF system for charging and controlling the IP traffic, it is useful to be able to interrupt a PDP Context session. In order to do this, the PCRF will have a proprietary interface with the AAA server, and a Disconnect-Request will be sent by the AAA server as below.

```
_____RECEIVED FROM PCRF (proprietary interface AAA server<-> PCRF___
Disconnect IP address  172.17.10.229
The AAA sever sends to the GGSN
_____SENT____ Header IP SENT _____
Src IP '192.168.2.2'
Dst IP '192.168.0.4'
- - - - Super Detailed GTP, SIP, DIAMETER, RADIUS Analyser (C)HALYS  - - message length = 60
____ Header RADIUS (port 3799): SENT ____
RADIUS Command: (40):Disconnect-Request
Packet identifier B3
headerRADIUS(with Authenticator)+payload = 60
Authenticator ABCDEFF101234567890123456789ABCD (MD5 calculated using shared secret:
totoalaplage)
 (26):Vendor-Specific(3GPP TS 29.061)
Vendor_ID = 10415 ((1):IMSI of handset)
value = 274040199000094
(44):Acct-Session-Id(Call ID of the session)
length AVP = 11
value = c2902093ee47102d
(8):Framed-IP-Address(IP of handset assigned by DHCP);presence means Terminal is IP
ACTIVATED!
length AVP = 6
value = 172.17.10.229
_____RECEIVED Header IP RECEIVED _____
Src IP '192.168.0.4'
Dst IP '192.168.2.2'
Port SRC 1813
String for MD5: 29B30014ABCDEFF101234567890123456789ABCD746F746F616C61706C616765
------Authenticator control MD5 of incoming RADIUS message---------
Received Authenticator 2075D045B26F0EBC6A6DD983796D0200
Sent Authenticator ABCDEFF101234567890123456789ABCD
Expected Computed received Authenticator 2075D045B26F0EBC6A6DD983796D0200
resultat validite Response Authenticator 0 Radius_code = 41
- - - - Super Detailed GTP, SIP, DIAMETER, RADIUS Analyser (C)HALYS - - - - - - - message length
= 20
____ Header RADIUS (port 3799): RECEIVED ____
RADIUS Command: (41):Disconnect-Ack
Packet identifier B3
headerRADIUS(with Authenticator)+payload = 20
```

Authenticator 2075D045B26F0EBC6A6DD983796D0200 (MD5 calculated using shared secret: totoalaplage)

4.2 National and International IP Roaming: Chain of RADIUS Servers in an IMS System

WiFi roaming allows a user which has a WiFi subscription attached to his mobile to use WiFi in another network. In the visited Access Point (hot spot), the RADIUS IP address is that of the visited WiFi. If the Access-Request is for halys.fr or wlan.mnc002.mcc623.3gppnetwork.org (EAP-SIM case), this is found in the body of the message. The visited RADIUS server must have a routing table of the networks it has an agreement with and resend (using a context) all the RADIUS messages to the home RADIUS. This does not involve as in the RCS application to resolve a MSISDN to find the domain.

4.3 Advanced Policy Charging and Control(PCC) Using a RADIUS Server

4.3.1 Standard Architecture

The charging and control of the data traffic is usually done in the HPLMN ("home routed access"). The architecture [4.1] is given in Figure 4.2 and may integrate the RADIUS server.

In the case of Figure 4.2, the RADIUS server is allocating the EndUserAddress (IP-PDP) to the TE in the Create PFP Context response. It has either its own IP pool or queries an external DHCP server. (1) is the creation of the PDP Context which is handled by the GGSN/PGW. A Radius Access-Request (2) is performed by the GGSN which passes MSISDN, APN and IMSI (*the only secured user identity*) as well as SGSN IP, IMSI, Cell ID (Circuit domain based), all useful for detailed charging.

The RADIUS server is, in this case, allocating the "Framed IP Address" IP Src from its internal pool using Access-Accept. Then, the GGSN starts the accounting session with a RADIUS Accounting-Request (Start) (3) which contains the allocated IP Src. The RADIUS Server is relaying (4) this Accounting-Request to the PCEF so that it knows that there is a new session. In (5), we see the Internet traffic, the IP Src is that of the User Equipment allocated by the GGSN in the Create PDP Response. The PCEF knows from the IP Src that this is a new session and requests (6) instructions from the PCRF with the IP Src using the Diameter Credit

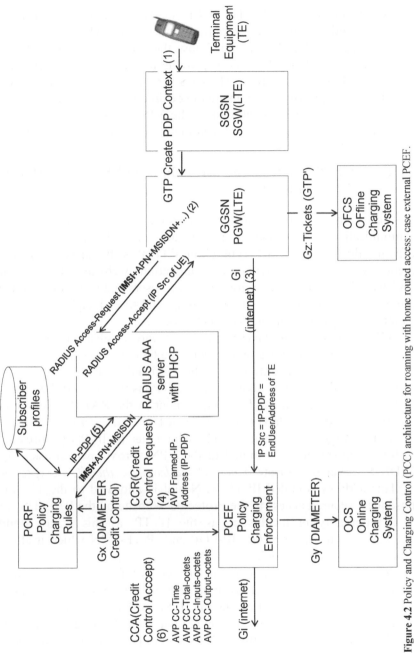

Figure 4.2 Policy and Charging Control (PCC) architecture for roaming with home routed access: case external PCEF.

Control Application [4.2]. The PCEF sends a Credit Control Request (CCR) (6) and will receive a Credit Control Acknowledgements (CCA). At this stage, the PCRF does not know the APN or the MSISDN, so it requests (7) the RADIUS server to provide them, so he can decide whom to charge and the QoS to apply. The QoS is returned (8) to the PCEF which can allocate the priority and manage the desired QoS.

Note that the Local Break-Out in the VPLMN is desired in the LTE specs in order to improve the QoS and the diagram is different in this case.

4.3.2 Secured Architecture with Ciphered End-to-End Tunnel and "Cryptophones"

This is a more sophisticated architecture [4.2] designed to secure communications for official or private services. The TEs have a client software (this makes them "Cryptophones") that is able to establish an end-to-end secured tunnel with the application website which contains confidential information only for authorized users and their communication cannot be usefully tapped (it is strongly end-to-end ciphered) by radio scanners or by probing in the fixed networks.

A VPN between the GGSN and the secured website is added to the previous system, it includes two servers, the GGSN side (the VPN "proxy" in which the address is configured in the TEs) and the PCEF side.

Until (3) it is the same as above, the GGSN has transmitted the "Framed-IP-Address"= IP Src of the User Equipment to the RADIUS server in an Accounting-Request (Start) (3). *But the RADIUS does not relay it to the PCEF* as in Figure 4.2. Then, the TE (which has the IP-PDP address) asks (4) the establishment of a tunnel to the VPN "proxy" which asks (5) the RADIUS server to allocate *a new IP address for the TE's cryptophone.*

This address is called IP-TE-VPN and is allocated to the TE for the subsequent secured access to the Protected Website by the RADIUS server (5) a proprietary protocol such as the HTTP, as this is not standard RADIUS which *will correlate* IP-PDP and IP-TE-VPN which it allocates from the same address-pool.

This *IP-TE-VPN is then provided in an Accounting-Request(Start)(6)* sent by the RADIUS to the PCEF to indicate a new IP session which uses IP-TE-VPN as Source IP. That means the IP Src observed by the PCEF is IP-TE-VPN *and not IP-PDP* returned in (2) when the TE accesses the website (7). This is necessary as the PCEF would have no way to know to recognize the beginning of a new session.

When asking instructions to the PCRF, the PCEF uses IP-TE-VPN (the IP Src address that it sees). The PCRF (9) as (5) in Figure 4.1

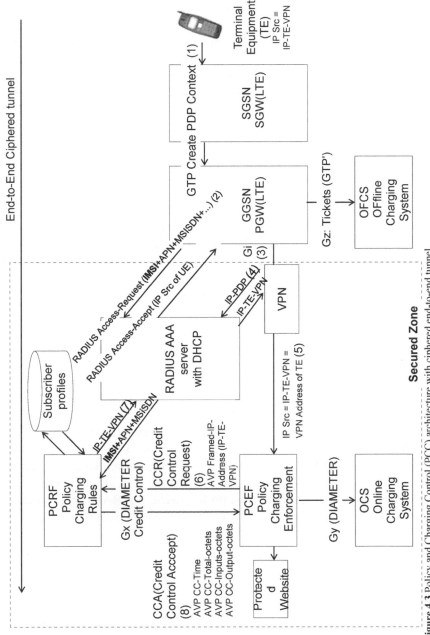

Figure 4.3 Policy and Charging Control (PCC) architecture with ciphered end-to-end tunnel.

interrogates the RADIUS server to know which user it is, but using IP-TE-VPN. From IP-TE-VPN and the correlation established (4), RADIUS provides APN, IMSI, MSISDN similar to the non-secured architecture, and the PCRF (10) provides the traffic handling parameters to the PCEF similar to (6) of Figure 4.1.

When the session is terminated by the UE (not shown on Figure 4.3), a RADIUS Accounting-Request (Stop) is sent by the GGSN to the RADIUS, which is also relayed to the PCEF.

Also not shown is the case where the PCRF decides that the session is inactive (no traffic) and sends to the RADIUS a proprietary "Disconnect". The RADIUS sends a RADIUS Diconnect-Request to the GGSN to free the context as in Section 4.1.4.

To further secure and provide the EAL5+level (this is one of the highest level), a SIM card with a SIM Tool Kit which implements the ciphering may be added to the design (some SIM card vendors provide it, using a certified cryptographic component). Only users who have the SIM card can then establish the secured connection.

We have explained the basic standard architecture, for simplicity the PCEF and PCRF may easily be integrated.

4.3.3 Using a GTP Hub-based GGSN: Added Simplicity for Service Provisioning

A standard GGSN is configured with service parameters defined for each APN that it handles:
If an APN is configured to use a RADIUS Access-Request, it will select the optional parameters in the list of Create PDP Context request parameters provided in the RADIUS Authentication. RADIUS Access-Request is not always used as the UPDATE LOCATION GPRS has performed the authentication based on the IMSI, but RADIUS Accounting always is. Most often APN, IMSI (mandatory to secure the identity), MSISDN, IMEISV, IP address of SGSN, Cell-ID that allow to have tariffs or QoS dependent on the location are selected to be included in an Access-Request

In this case, the QoS (priority) is defined by different APNs which need to be changed in the TE and in the HLR for each user.

A flexible GGSN, in particular, based on a GTP Hub, will be able to ignore the need for a configured APN. It will allow the service based only on the IMSI contained in the Create PDP Context and will let the PCRF decide entirely the QoS. No change of APN will be necessary including in the HLR if the wildcard '*' is used as APN to bypass the control performed by the SGSNs (comparison between the APN of the TE and the list provided by the HLR). The GGSN may be included in the "secured zone", also it makes tapping attempts more difficult although the end-to-end tunnel architecture should make this ineffective.

In the case that the service uses multi-IMSIs, the IMSI passed to RADIUS (2) will be 'auxiliary' not 'nominal' for a standard GGSN. On the contrary, a GTP Hub would always provide 'IMSI nominal' and is then simpler for service provisioning in the PCRF. If there are several auxiliary IMSIs it would become quite complicated because the PCRF would have to handle the correspondence table IMSI auxiliary(s)->IMSI nominal which is not his function. The MSISDN which is provided will the same irrespective of whichever IMSI auxiliary is used, but it is not secured as the IMSI which is checked by the AuC in the HLR which has strict security conditions.

For such a secured architecture able to support multi-IMSI virtual data roaming, a GTP Hub-based GGSN is almost mandatory in providing the IMSI nominal identity to the PCRF via a RADIUS server in step (2) which is not an auxiliary IMSI. In order to have the maximum security (without any possibility to change the correspondence table), it is another reason to have the GGSN/GTP Hub in the "secured zone".

References and Further Reading

[4.1]   3GPP TS 29.215 V11.7.0 (2013-1), "Technical Specification Group Core Network and Terminals: Policy and Charging Control (PCC) over S9 reference point: Stage 3 (Release 11)"
[4.2]   RFC 2865, "Remote Authentication Dial In User Service(RADIUS)",June 2000 (RADIUS on port 1812)
[4.3]   RFC 2866, "Radius Accounting", June 2000 (RADIUS on port 1813)
[4.4]   RFC 2867, "Radius Tunnel Accounting",June 2000 (extensions of RFC 2866).
[4.5]   RFC 5176, "Dynamic Authorisation Extension to RADIUS", June 2008 (RADIUS on port 3799)
[4.6]   RFC 4006, "Diameter Credit Control Application", August 2005

5

SIP Hubs and RCS Hubs (Rich Communication Suite)

One does not easily deceive the watchfulness of comrade Stalin

  −Joseph Vissarionovitch Dougachvili

5.1 Issues and Practical Implementation of the Mobile Address Resolution (MARS)

The basic issue related to MNP has been going for more than 10 years with a strong pressure of some service vendors to implement a central database (the ENUM database) that they would be selected to provide. Some of their proposed opex charging schemes such as charging for each MARS request lead to search for alternative solutions. Another reason is that many countries are unwilling for legal privacy protection or for national security reasons, to provide to a third party the MNO name of each of their mobile numbers.

Example of an ENUM request: Requested number +33508123456. It is translated [5.1] as

6543218033.e164.arpa

The DNS request to the ENUM server returns a set of NAPTR records (for VoIP, email) which includes the IP address of the destination IMS server.

The most flexible implementation of the M-DNS is a DNS software. From the RCS Hub, it is interrogated with a C language *gethostbyname.* However, it can also be used, instead of the local DNS resolver by the MNOs' IMS, providing an additional service of the RCS Hub for the national interco. The RCS traffic flow and the various address resolution method are illustrated by Figure 5.1.

117

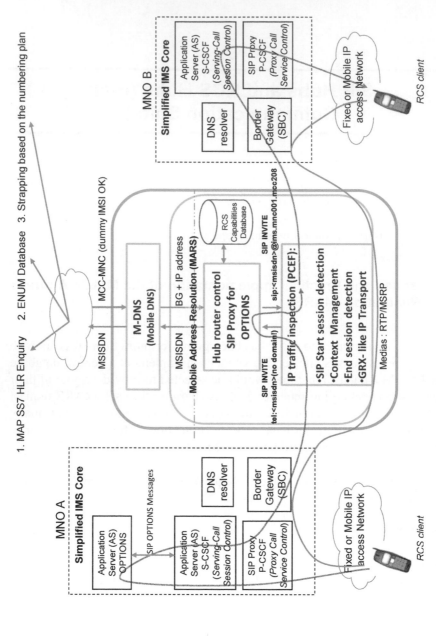

1. MAP SS7 HLR Enquiry   2. ENUM Database   3. Strapping based on the numbering plan

MNO B

Simplified IMS Core

Application Server (AS)
S-CSCF
(*Serving-Call Session Control*)

DNS resolver

SIP Proxy
P-CSCF
(*Proxy Call Service Control*)

Border Gateway (SBC)

Fixed or Mobile IP access Network

RCS client

MCC-MNC (dummy IMSI OK)

RCS Capabilities Database

SIP INVITE
sip:<msisdn>@ims.mnc001.mcc208

M-DNS
(Mobile DNS)

BG + IP address

Mobile Address Resolution (MARS)

Hub router control
SIP Proxy for
OPTIONS

IP traffic inspection (PCEF):

•**SIP Start session detection**
•**Context Management**
•**End session detection**
•**GRX- like IP Transport**

MSISDN

MSISDN

SIP INVITE
tel:<msisdn>(no domain)

Medias : RTP/MSRP

MNO A

Simplified IMS Core

Application Server (AS)
OPTIONS

SIP OPTIONS Messages

Application Server (AS)
S-CSCF
(*Serving-Call Session Control*)

DNS resolver

Border Gateway (SBC)

SIP Proxy
P-CSCF
(*Proxy Call Service Control*)

Fixed or Mobile IP access Network

RCS client

Figure 5.1 RCS Interco HUB with SIP OPTIONS traffic trimming.

5.2 Optimizing the SIP OPTIONS Traffic: Use of a SIP Proxy "Options"

One needs to compare the use of the OPTIONS SIP message between a traditional SIP service and RCS. OPTIONS is used to interrogate the capability of a distant client. Without any precaution, there would be one MARS interrogation each time an OPTION is sent.

The OPTIONS message traffic is then quite heavy in RCS compared to SIP and it is desirable to trim this traffic with the introduction of a specialized Application Server in the IMS architecture, to be called AS OPTIONS (to optimize the flow for the national RCS service) or in the RCS interco Hub for the RCS service between different countries.

Note that there have been proposals that the central ENUM database would also hold a customer profile with their RCS capabilities and others. The privacy protection or the security obligations defeat this naive approach, and the described RCS Interco Hub with the "OPTIONS Proxy" is more realistic.

5.2.1 Architecture of the SIP Proxy for OPTIONS

It includes a RCS capabilities database. It is refreshed by the successful RCS OPTIONS subscribers of MNO B to subscribers of MNO A as well as by the periodical refresh policy explained below. Table 5.1 provides the differences between the RCS and SIP protocols.

Table 5.1 Comparison between the RCS and SIP protocols

RCS	*SIP*
An OPTIONS message is sent after a successful REGISTER to all contacts of the address book	*An OPTIONS message is sent after a successful REGISTER to all contacts of the address book*
Making a call (INVITE) using VoLTE *An OPTIONS message is sent to the destination client to check that his capabilities have not changed.*	*Making a call (INVITE) or sending a MESSAGE* *No OPTIONS message sent*
Browsing through one's address book while idle. *An OPTIONS message is sent to each contact*	*Browsing through one's address book while idle.* *No OPTIONS message sent*

5.2.2   Optimizing the Periodical Refresh of the OPTIONS of the Address Book Contacts

The RCS Capabilities of a particular destination correspondent **B** of a subscriber **A** of MNO **A** will be refreshed, when it tries a RCS session to subscriber **A**. This is an automatic source of refresh.

Until he does that, **A** cannot RCS with him because the OPTIONS sending is blocked. It is then useful to have periodical refresh attempts where OPTIONS is sent to **B** every T period even if the previous attempts have shown that he is not RCS-capable. He may have become capable since the last attempt. The interval T must be optimized with the cost O of a MARS request and the revenue C of the RCS service of **A** with **B** (individual charging, monthly fee, etc.).
Let:

λ for a given subscriber **B**, the rate of his initial subscription to the RCS service (number / month)

T the interval since the last OPTIONS message was sent to **B**, and a new OPTIONS is sent

C the revenue to the MNO of a if a new RCS correspondent **B** is added to the RCS client **A**

O the cost of an OPTIONS message (ENUM request, MAP SS7 traffic, etc.)

One finds that the optimal T is:

$$T = ((Ln(1 - O/C))/ \lambda, \text{ with } O < C \tag{5.1}$$

Proof:
When subscriber **A** is attempting to contact the RCS correspondent in his address book, for a given contact **B**, the "Proxy OPTIONS" can decide:

Not to send OPTIONS to **B**: the net revenue is R= C – O, both = 0,

To refresh the OPTIONS for **B**: Assuming the usual exponential Markov distribution of intervals between subscription, the probability of success is $e^{-\lambda T}$. And, the expectation of the net revenue is:

$$R = C(1 - e^{-\lambda T}) - O,$$

Which is $>=0$ if the refresh attempt is made at time $t >= T$ given by [5.1]

Numerical examples: $C/O = 5$ (the revenue is more than the cost, otherwise there is no interest to offer the RCS service); $\lambda = 0.1/$ month (the subscribers take 10 months in average to decode to subscribe to RCS). [5.1] gives $T = 2.23$ months. If $C/O = 20$ (the MARS requests are cheap), it gives $T = 0.513$ month, the "OPTIONS proxy" will attempt to refresh the OPTIONS of the correspondent of subscriber A, every half month.

Further Readings

[5.1]   RFC 6116 The E.164 to Uniform Resource Identifiers (URI), Dynamic Delegation Discovery System (DDDS) Application (ENUM).

6

Diameter Hubs (LTE virtual roaming Hubs)

6.1 Charging with Diameter

DIAMETER (double RADIUS was the reason for the name) was introduced to replace RADIUS with more possibilities and security. It uses TCP (connected) instead of UDP. The specification includes several applications, the first being "Charging" which we will start with:

Voice call charging,
SMS charging.

6.1.1 Data Traffic Charging

With the introduction of LTE (4G), it has been chosen as the full replacement of MAP which used ASN1 message coding, TCAP , SCCP. It loses the flexibility of TCAP dialogues which was mainly necessary to help the segmentation of long messages due to the limited size of SCCP messages (256 octets). DIAMETER is only request-answer but uses IP packets for transmission which are much longer (1400 octets about).

6.1.2 Voice Call Charging

This concerns the real-time charging for pre-paid customers. Refer to [6.2] for the CAMEL protocol and usage which is between the MSC/VLR/SSF and the SCF (System Control Function) part of the IN system. This is assumed to have a canonical architecture with a SDP (System Data Point) which is the rating engine having the subscriber's credit and their tariff plan. DIAMETER (Credit Control Application) is used between the SCF and the SDP. The classical call flow for Camel voice calls charging is given by Figure 6.1.

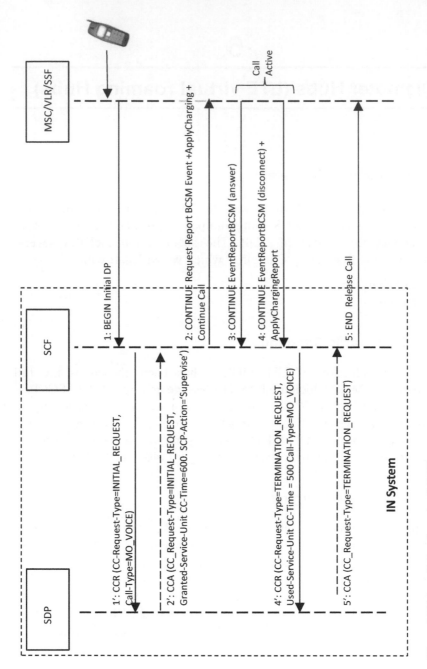

Figure 6.1 Voice call charging using CAMEL.

6.2 LTE Roaming with S6 Diameter Based Protocol

There is a major simplification difference between SS7 roaming Hubs mono-IMSI and multi-IMSI [0.9] and LTE Roaming Hubs (as well as GTP Hubs). SS7 has two network layers and TCAP.
DIAMETER and GTP have only the IP network and no TCAP. To link Requests and Responses they use a Session-Id (LTE case) or a Tunnel End Point Identifier (GTP case). Here is the content of a LTE UPDATE-LOCATION Request (equivalent to UPDATE-LOCATION-GPRS Request with SS7.

< Update-Location-Request>
  < Session-Id >

  { Auth-Session-State }
  { Origin-Host }
  { Origin-Realm }
  [Destination-Host]
  { Destination-Realm }
  { User-Name }
  *[Supported-Features]
  [Terminal-Information]
  { RAT-Type }
  { ULR-Flags }
  [UE-SRVCC-Capability]
  { Visited-PLMN-Id }
  [SGSN-Number]
  [Homogeneous-Support-of-IMS-Voice-Over-PS-Sessions]
  [GMLC-Address]
  *[Active-APN]
  [Equivalent-PLMN-List]
  [MME-Number-for-MT-SMS]
  [SMS-Only]
  [SMS-Reqister-Request]
  *[AVP]
  *[Proxy-Info]
  *[Route-Record

If the Request is relayed from a visited network to a HPLMN by a LTE Hub, the Response of the HPLMN is always routed back to the IP Source address (of the LTE Hub). This is not the case for SS7 Roaming Hubs, where the Response is routed back to the SCCP Calling Party Address of the Request (see [0.8] Figure 5.1). In order to have it back to the Roaming Hub, in the case the connection is not direct using "direct MTP", the "Alias GT" system was invented. The Roaming Hub replaces the original SCCP Calling Party Address by one of its addresses.

References and Further Readings

[6.1] 3GPP TS 32.299 V.11.5.0 (2012-09), "Diameter Charging Applications , Release 9"

[6.2] R.Noldus,"CAMEL, Intelligent Network for the GSM, GPRS and UMTS Network",Wiley, 2006 (an excellent book describing all the CAMEL architectures).

7

MMS Hubs

This chapter explains the implementation of a MMS Hub for developers and operation staffs, but it is not sufficient to develop MMSC as several other protocols are needed, in particular MM1 which handles the transfers with the handsets. An MMS Hub needs only the MM4 protocol. The MM4 protocol uses the SMTP transport layer (the standard mail exchange protocol). A full explanation of the MMS interfaces and protocol is given by [7.1].

7.1 MMS Hub Principle and the MM4 Protocol

The graph below represents the MMSCs of the MNOs and the MMS Hubs which communicate through the GRX network. Each has their own IP address. The MMS Hub providers (Aicent, France Telecom, Vodafone, Belgacom, etc.) make "peering agreements" which allow to share the MNO destinations they have a direct contract with in order to provide a better global coverage. The graph of Figure 7.1 represents such a MMS network. The GTP Hubs do not handle directly the MMS with the handsets (MM1 protocol) or with various Value Added Services (VAS with the MM7 protocol), so they implement only the MM4 protocol used to relay the various MMS messages with their peering partners and with their client's MMSCs. All the MMS protocols are described in [7.1]. The MM4 protocol uses the SMTP standard mail protocol, with a MMS header and the content coded in the MIME standard (pictures, sound, etc.)
One can see that Glomobile (Nigeria) has its own direct MMS agreements, as well as a contract with one of the GTP Hubs. Some of the direct MMS destinations could be also reached by a peering partner, such as MMS Hub#2, of their GTP Hub providers as there are multiple paths in the graph. It is up to the MNO's MMSC to use the shortest route which will be a direct agreement when they have it, in order to reduce their MMS termination cost. For example, Glomobile would send the MMS direct to MTN Nigeria.

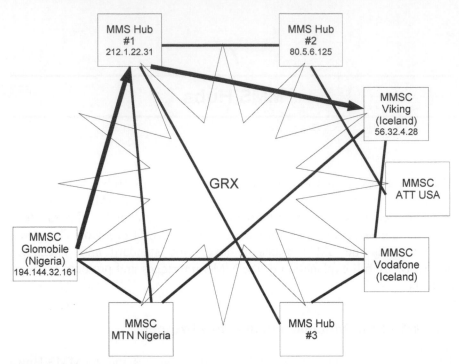

Figure 7.1 Graph of the MMS Hub network.

The call flow of Figure 7.2 shows the various messages of the MM4 protocol in the most complex case when the sending handset asked a "Delivery Report" upon MMS reaching the mobile destination and a "Read Report" when the MMS is read by the destination customer, which could be much later. This Read Report is something which does not exist in the SMS service. Between the handsets and their MMSC, two transport methods are used:

- HTTP (POST to send a message and GET to retrieve the MMS)
- SMS (a MM1 m-notify-ind (known as "MMS notification SMS") is sent to the destination handset to provide the URL to retrieve the MMS). The handset on the right can retrieve this MMS with a HTTP GET request

It then sends a m-notify-ind notification to his MMSC (Viking) that it has received the MMS using a HTTP POST. When the MMS is opened by the customer, a m-read-rec-ind is sent with a HTTP POST to inform the sender that the MMS has been read.

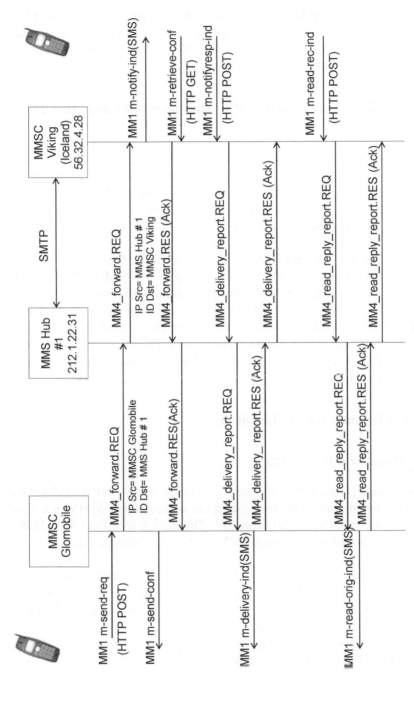

Figure 7.2 Call Flow of the MMS service in the MMS Hub network.

As we are interested in MMS Hubs we will give details only of MM4 as this is the only protocol they need to have. As you can see, six MM4 messages will be received and sent by MMS Hub#1, which needs an efficient mailer system.

7.2 MMS Transport Service: the Standard Mail with the SMTP Protocol

A trace in a MMS Hub#1 of the MM4_forward.REQ is the best way to understand, what is the MM4 envelop, how the content (a picture in the example) is coded as a MIME[7.2]. The trace shows the result of the routing algorithm which determines the IP address of the next MMS Hub or the final MMSC and the sending of the MM4_forward.REQ using the SMTP mail protocol.

First, the MM4_forward.REQ received from the MMSC of a customer. Then, the MM4_forward.REQ relayed to the peer MMS Hub# 2 as there is no direct agreement with the final destination network. The resolution of the destination number +447718123456 uses an interrogation of the HLRs with the SS7 network. If it fails the "strapping" method is used to determine the destination network from a comprehensive numbering plan. If there is MNP in the destination country this may cause a delivery failure, but there is no other way.

7.3 MMS Domain Name Resolution

7.3.1 Using the Root DNS

How is Section 1.3 applied to the case of sending MMS from a MMSC or a MMS Hub to another? The base transport service for a MMS is a mail which uses the standard SMTP protocol. The mail is sent from the MMSC Glomobile to the subscriber distant MMSC Viking. To send a mail:

- the distant network must be "resolved", that is, finding the network name from the MSISDN (e.g. Viking: +35465123456. This is done by interrogating the HLR with this MSISDN which gives the IMSI which belongs to Viking,
- the MCC-MGT is derived from a table : 621-50 and a mms domain name is built mms.mnc050.mcc621.gprs,

- the local DNS of the sending MMSC resolves recursively with the use of the root DNS address. In general, *the used DNS is a replicate of the main root DNS*, replicate provided by the GRX provider of the sender.

An important remark is that the DNS of the GRX is not properly fully provisioned. A MMSC or a MMS Hub should provision its own local DNS with the MMSC IP addresses of all their partners, and configure the DNS search as follows:

Primary DNS = local DNS
Secondary DNS = GRX DNS (if a domain is missing in the local DNS)

To check that the DNS resolution is properly set in a UNIX system *connected to the GRX network* (of course) use the "dig" command. In the example, we use the optional ANY parameter in order to have also the optional AUTHORITY SECTION and the ADDITIONAL SECTION with the addresses of the two Root DNS used in the recursive procedure because mnc050.mcc621.gprs is not found in the local DNS.

dig @localhost mms.mnc050.mcc621.gprs MX ANY

;; flags: qr aa rd ra; QUERY: 1, ANSWER: 1, AUTHORITY: 2, ADDITIONAL: 2
;; QUESTION SECTION:
;mms.mnc050.mcc621.gprs. IN ANY
;; ANSWER SECTION:
mms.mnc050.mcc621.gprs. 86400 IN A 80.255.61.23   /* gives the IP of the destination MMSC */

;; AUTHORITY SECTION:
mnc050.mcc621.gprs. 86400 IN NS grxns1.opentransit.grx.
mnc050.mcc621.gprs. 86400 IN NS grxns2.opentransit.grx.
;; ADDITIONAL SECTION:
grxns1.opentransit.grx. 86400 IN A 193.251.244.130 /* Address of DNS root replicate
grxns2.opentransit.grx. 86400 IN A 193.251.244.162 /* other Address of DNS root replicate

The MMSC will send the MMS using the MM4 protocol to the MMSC of Glomobile in which the IP address 80.255.61.23 has been returned by the DNS procedure.

7.3.2   More General: Populate the Local DNS

When a MMSC or MMS Hub is operated, one finds that many distant partners do not enter properly all the APNs in their local DNS which will be interrogated recursively by the root DNS. In many instances, the local DNS of the sending system must be manually populated with the IP of the MMSCs of the partners.

7.3.3   Routing Algorithm in the MMS Hubs

7.3.3.1   Trace of the Reception of a MMS

The MMS is first received by the standard mailer of the system which uses the SMTP protocol. It provides the MAIL FROM parameter to the MMS main process, as well as the **Two** RCPT TO which identifies a destination mail address. In case the sender sends the same mail to several destinations *there would be multiple RCPT TO addresses* and the sender MMSC could group the sending in a single mail. This is rarely used by MMSCs, they send several mails when they need to send to multiple destinations in most cases. To explain the situation, we take the case of sender MMSC grouping the MMS.

The trace in the MMS main process includes:

Parameters obtained from the mailer (example of a mail sent to a single destination)

MAIL FROM:<+2348051234567/TYPE=PLMN@mms.gloworld.com>
RCPT TO :   +3548993254/TYPE=PLMN@mmshub1.fr *// RCPT TO parameter # 1*
RCPT TO :   +3548968466/TYPE=PLMN@mmshub1.fr *// RCPT TO parameter # 2*

Parameters in the MM4 header

Return-Path: <+2348051234567/TYPE=PLMN@mms.gloworld.com>
Received: from Glomobile_mm4 (mms.mnc002.mcc274.gprs [194.144.32.161]) by mmshub1.fr (Postfix) with ESMTP id ED818246734   for <+3548993254/TYPE=PLMN@mms.imc.is>;
Message-ID: <32248757.1373065735231.JavaMail.root@ice-fe21>
Date: Fri, 5 Jul 2013 23:08:50 +0000 (GMT)
From: +2348051234567/TYPE=PLMN
Sender: +2348051234567/TYPE=PLMN@mms.gloworld.com
To:+3548993254/TYPE=PLMN,+3546601701/TYPE=PLMN,+3548409122/TYPE=PLMN,+3548400057/TYPE=PLMN,+3548968466/TYPE=PLMN,+3548972209/TYPE=PLMN
X-Mms-Originator-System: mm4@mms.gloworld.com

The body of the MMS including the various parts coded in MIME

Mime-Version: 1.0
Content-Type: multipart/related;   type="application/smil";   boundary="----
=_Part_26494_10783063.1373065735215"; start="<smil>"
X-Mms-Message-Type: ***MM4_forward.REQ***
X-Mms-Transaction-ID: "bCo1ZCCUWHQB3"
X-Mms-3GPP-MMS-Version: 5.2.0
X-Mms-Message-ID: "XI14aCmW2PQB3"
X-Mms-Read-Reply: No
X-Mms-Sender-Visibility: Show
X-Mms-Message-Class: Personal
X-Mms-Priority: Normal
X-Mms-Delivery-Report: No
X-Mms-Expiry: 604795
X-Mms-Ack-Request: **No /* in this example the sender MMSC does not ask for a MM4_forward.RES */**
Content-Type: [88]vnd.wap.; start: ; type: ; boundary:
------=_Part_26494_10783063.1373065735215
Content-Type: application/smil; name=smil.xml
Content-Transfer-Encoding: 7bit
Content-ID: <smil>
Content-Location: smil.xml   *// the smil part is necessary to organize the presentation of the various parts of the MMS body.*
<smil>
  <head>
    <layout><root-layout width="1230px" height="720px"/>
    <region   id="Text"   left="0"   top="648"   width="1230px"   height="72px"
fit="meet"/><region   id="Image"   left="0"   top="0"   width="1230px"   height="648px"
fit="meet"/>
    </layout>
  </head>
  <body>
    <par   dur="5000ms"><text   src="cid:text_0.txt"   region="Text"/><img
src="20130705_225716.jpeg" region="Image"/></par>
  </body>
</smil>
------=_Part_26494_10783063.1373065735215
Content-Type: image/jpeg; name=20130705_225716.jpeg
Content-Transfer-Encoding: base64
Content-ID: <20130705_225716>
Content-Location: 20130705_225716.jpeg
......peg image coded in Base 64

AQEBAQEBAQEBAQEBAQEBAQEBAQEBAQEBAQEBAQH/2wBDAQEBAQ
EBAQEBAQEBAQEBAQEB
AQEBAQEABAQEBAQEBAQEBAQEBAQEBAQEBAQEBAQEBAQEBA
QEBAQH/wAARCAJEAbMDASIA
....

When the MMS is sent by the MMS Hub, the Body remains identical (there is no resizing of the images, which is the destination MMSC responsibility if needed). The four parameters which are changed by the MMS Hub according to rules are shown with bold characters. They concern all the parameters which have the domain name of the sender; this is adjusted according to the setup agreed with the destination network or a peering MMS Hub, such as MMS Hub#2. For a given originating network, two types of domain names can be used (example Glomobile):

- domain name(DN) GPRS: mms.mnc050.mcc621.gprs,
- domain name(DN) internet: mms.gloworld.com.

These transformations apply to the six MM4 messages of Figure 7.2 except X-Mms-Originator-System *which is only in the MM4_forward.REQ.*

7.3.3.2   Multiple Destinations Handling: The MMS Hub Does Not Perform Any Grouping

The example shows that there are six Iceland (+354) destinations but only **two** +3548993254 and +3548968466 have a RCPT TO in the received MMS.
The MMS Hub creates a list of the To: , Cc: and CCi: *but sends mail only to those which are in the RCPT TO list of destination numbers*. The mailer system must *be configured to provide these SMTP parameters* to the MMS Hub software.

The recipient handset will see all the destinations in copy. This is necessary when one looks at Figure 7.1 as Glomobile has a direct MMS agreement with the Vodafone Ireland to send a mail to each destination number in general (no grouping). If the MMS Hub#1 had sent MMS to the other To: they would have revived the MMS several times.

The MMS Hub does not do any grouping and will send two separate MMS to +3548993254 and +3548968466.

7.3.3.3   Address Resolution

This is again a very important and delicate issue as MNP is very frequent including in Iceland. The mobile destination +3548993254 is a Viking subscriber ported in from Vodafone. The MMS Hub#1 must recognize it is a Viking number and set the proper domain name for sending.

The most practical method id to use the HLRs interrogation (this is called the MM5 interface of a MMSC or MMS Hub). The HLR returns the IMSI which allows to find the DN GPRS or the DN internet with a numbering plan table. In case of failure, most MMS Hubs use "strapping", they decide the network name based only on the number +3548993254, this may cause failures of course.

7.3.3.4 Shortest Path Algorithm, Default Transformation Rule and Sending

Once the destination network name and the domain has been determined by MM5 or "strapping", the best method to determine the routing is to use a fast shorter path algorithm such as in [0.5]. The provisioning of each MMS Hub allows us to represent the graph of his adjacent destinations and they must have the list of destinations of their MMS Hub peers.

MMS Hub#1 finds a direct route with Viking, but for ATT (USA) the shortest path starts with MMS Hub#2.

In order to avoid to configure for each destination the table 7.1, it is better to have a default transformation for each direct destination and each adjacent peer. A particular entry is created only if needed and not systematically for each destination.

Example of default transformation for the MMS sent to (RCPT TO) +3548993254 ;
MAIL FROM: User transparent@DN MMS Hub#1
RCPT TO: User Transparent@DN GPRS or User Transparent@DN adjacent Hub (routing to a Hub)
Sender: Copy MAIL FROM
X-Mms-Originator-System: User Hub (MMSC) with DN MMS Hub#1

Table 7.1 Table of domain name transformations

SMTP or MM4 parameter name	Possible values configured by the MMS Hub#1
MAIL FROM	Full transparent (user@domain) User transparent@DN internet User transparent@DN GPRS User transparent@BN MMS Hub#1
RCPT TO	User transparent@DN internet User transparent@DN GPRS User transparent@DN adjacent Hub
Sender	Full transparent (user@domain) Copy MAIL FROM
X-Mms-Originator-System (this is used by the destination network to bill the originator for the termination of the MMS)	Full transparent or copy of MAIL FROM if not present. User Hub(MMSC) with DN internet User Hub(MMSC) with DN GPRS User Hub(MMSC) with DN MMS Hub#1

Once this is done, the MM4 is sent to the Viking MMSC in a mail which would have
FROM: +2348051234567@mmhub1.gprs
TO: +3548993254@mms.mnc004.mcc74.gprs

Another separate MMS will be sent by the MMS Hub to (RCPT TO) +3548968466

7.3.3.5   Billing

Following a good billing principle, the MMS Hub will charge only the message delivered to the next destination, other MMS Hubs or destination MMSC. It needs to receive a MM4_forward.RES, this is the reason why for the example of 7.3.3.1, it will force

 X-Mms-Ack-Request: **Yes.**

The MM4_forward.RES which it receives will not be relayed to Glomobile which did not request it. If the sending of the MM4_forward.REQ is delayed and retried by the mailer, the billing would be delayed until the MM4_forward.RES is effectively received.

7.4 Developing a MMSC or a MMS Hub

The readers who are tempted must know that this is a very significant undertaking, compared to developing a SMSC. There are several complicated protocols, even the correct parsing of MM4 is difficult (several lines for a given parameters such as To:). Open code source implementations do not provide address resolution and their compliance with the standards is approximate.

Further Readings

[7.1]   3GPP TS 23.140 v6.16.0(2009-4) "Universal Mobile Telecommunications System (UMTS);Multimedia Messaging Service (MMS);Functional description; Stage 2, Release 6"
[7.2]   RFC 2845, Multipurpose Internet Mail (MIME) Part One: Format of Internet Message Bodies (there are many other RFCs which concern the MIME format)

8

GSM<>IP Seamless number continuity

8.1 User Experience Service Description

Experience # 1: GSM in the aircraft but disabled
This was a real experience (2012) during a flight to Hong Kong, as well a flight over China where the GSM in-flight usage is ruled out. An air traveler was called from ground on his GSM number while he was registered on the aircraft's WiFi connected with a "MSC/VLR-SIP" ground system (which provides the Transparent Call or SMS TO GSM Number / VoiP conversion, see [0.8 chapter 10]).

The air traveler was using a PC to test the visio quality and also an iPhone, both using standard "SIP clients". The voice quality was quite good and even considered by the traveler as "as good as GSM". SMS were sent (and received) both from the ground and from the passenger. This in-flight usage shows that the internet access (which uses the cheaper satellite "background service") in modern aircrafts has the quality to provide in-flight voice and even visio services using IP. The regular "Sub Band" satellite service used for in-flight GSM (provided by ONAIR, Aeromobile, MegaFon, etc.) has a guaranteed quality of service and is more expensive.

This is a practical user case (no GSM coverage, the aircraft had only WiFi activated) of a VoIP<->GSM system, which is used since two years by several GSM operators to provide cheaper roaming for their outbound subscribers, and also recently to provide voice, visio and SMS services to private or business customers having their PC equipped with their 3G or 4G USB "dongles" for internet wireless connection.

Experience # 2: GSM in the aircraft but no agreement with the in-flight operator
A traveler from Somalia was flying with Oman Air (both GSM- and WiFi-enabled) from Masquat to Paris. His Golis Telecom GSM did not have the roaming agreement with the in-flight operator of Oman Air. He subscribed to the WiFi access with his credit card, and used it for VoIP and visio. The

in-flight operator operates also the Wifi and gets revenues coming from the direct credit card payment. Using a pre-paid charging IN system or a centralized RADIUS server in the in-flight operator's ground segment, the credit can be kept from one flight to the other. This very frequently occurring business case as the main in-flight operators have concurrent agreements with about one-third of all the GSMs (they are over 800).

Experience #3: no GSM in the aircraft
Business trip from Paris to Dubai. The traveler needed to be able to receive some important calls from his office in Paris, but Emirates had only WiFi on their A380 (mid-2012). The traveler connected his Android smartphone to the aircraft WiFi and received the calls he was waiting for.

Improved voice quality over GSM through cabin noise adaptive filtering
When one talks with an in-flight correspondent, the surrounding cabin noise is rather high and degrades the comfort. However, the noise is "stationary" (the spectrum of Figure 8.1 is stable for a long time).

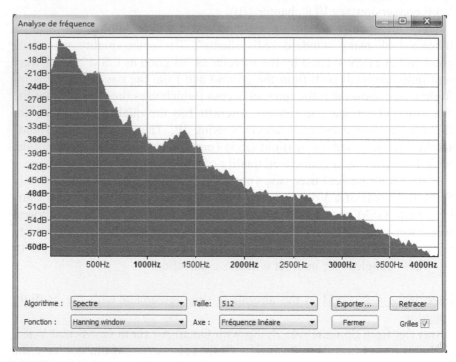

Figure 8.1 A380 cabin noise spectrum recorded in the ground segment GSM<->VoIP.

When using VoIP either from a standard PC or a smartphone SIP client, the voice flow (RTP packets) goes through the ground system which can make an adaptive filtering of the signal which uses "spectral subtraction": the estimate of the noise's spectrum is subtracted from the received speech + noise's spectrum.

8.2 Detailed Architecture

The principle and diagram is in Figure 8.2. The Hub used combines a SS7 MSC/VLR/SSF, a SIP server and a Media Gateway in a "pre-IMS" architecture.

8.3 WiFi Access Case

One wants to avoid the distribution of passwords to the WiFi access users. Since 2010, many smartphones implement the EAP-SIM access authentication method. This method can be used with Access Points 802.01X which implements the EAP authentication with a RADIUS server.

Since two years, smartphones and tablets support the EAP-SIM authentication for the WiFi access, which involves the SIM or USIM card in the initial authentication with the HLR. The GSM and WiFi operators (such as Orange France, SFR , Free, etc.) can then offer a seamless usage of all the data services, voice and messages when the user is WiFi-connected. This is most useful to offload the GPRS network, and for smaller operators, allows to cover small areas such as villages with inexpensive WiFi base stations with the same charging policy.

In [8.1], the use of SIM cards to authenticate users was already forecasted with USB card readers plugged to the PC and with smartphones, allowing to charge the customers using Voip, similar to GSM coverage. Today, the newly developed solution [8.2] permits to authenticate users via EAP-SIM protocol.

The system described fully supports this authentication scheme, standardized since 2008, with a AAA-MAP/SS7 Gateway, the data services in a SGSN gateway and the circuit mode VoIP-SMS with the Media Gateway, all three being integrated in the same platform. The need for a separate VoIP authentication is suppressed and also the need to distribute authentication keys. This is an important step to make the use of VoIP system by the GSM operators to become seamless with current 3G. The system described is the WIFI access subnetwork of a 4G LTE core.

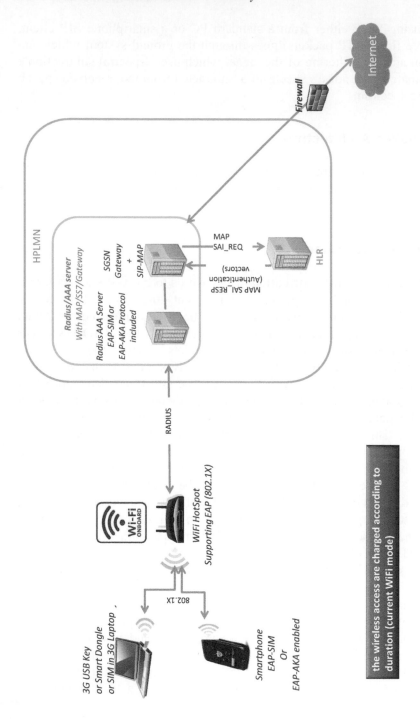

Figure 8.2 Entities involved in the EAP-SIM authentication process.

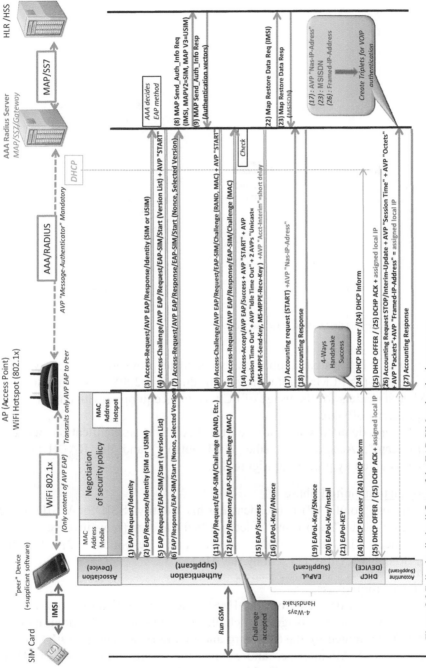

Figure 8.3 Fully detailed call flow of the EAP-SIM authentication process.

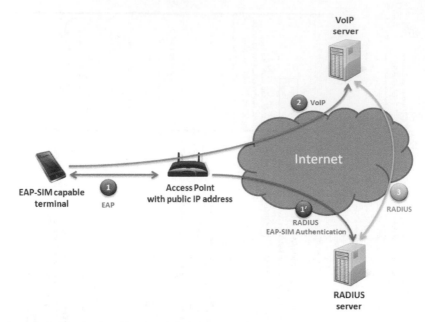

Figure 8.4 Activating the "Circuit mode" services.

The OTA SIM system should be extended to provision the new WLAN preferences (as PLMN preferences) introduced in 2008. One can use USB SIM card readers for the usage of PC terminals equipped with an EAP-SIM "supplicant software".

A very detailed Call Flow of the EAP-SIM authentication between "supplicant", Access Point (the WiFI hotspot) and the RADIUS-AAA server is given by preceding Figure 8.3.

"Association" initial phase
The Access Point (AP) is broadcasting his identity (SSID or WSID in the LTE case). The "supplicant" selects a WiFi access (may involve WLAN preferences in the USIM card) and responds (2) to the EAP/Request/Identity (1) sent by the AP.

Authentication phase
This identity is derived from the IMSI. The "Authentication" phase (5 to 17) involves the exchange of "challenges" based on Random number (RAND) computed by the SIM card and by the HLR. The RADIUS server integrates a MAP interface and obtains the "challenges" from the HLR with a MAP SEND_AUTHENTICATION.

There is a radio security between the "supplicant" and the AP using keys. The supplicant and the RADIUS server both compute a common "Pairwise Master Key" (PMK) based on the result of challenges with the SIM card on one side and the HLR on the other (Kc, SRES ...). At the end (16) of a successful Access-Accept, the RADIUS server sends them, ciphered, to the Access Point using their "shared secret". The AP is able to decipher this PMK (also called "Unicast") and using a "4-way handshake" with the supplicant will validate this key to be used for protecting their exchanges. After this, the AAA-RADIUS server obtains the MSISDN with a MAP RESTORE_DATA (25–26).

   There are many security signatures and this makes the implementation of RADIUS software for EAP-SIM or EAP-AKA a rather delicate project.

IP address allocation to the device
The "Device"(the Terminal not the supplicant software) then obtains a local IP address from the "DHCP" steps 22–23.

Accounting phase
It starts at (26). The "Accounting-Request" includes the local IP address assigned to the device, the same Accounting-session-Id as used previously, and the AP IP address. This allows the AAA-RADIUS server to find the previously obtained MSISDN which, with the local IP address and the AP IP is used as an authentified "triplet" for the usage of the VoIP services.

   Local IP Address of the Terminal-Access Point IP address-MSISDN

At the end of this successful process, the customer can now access all Internet services.

Using the EAP-SIM applet in the USIM card
There are free software supplicants (XS supplicant, WAP, etc.) available which supports EAP-SIM.

   To simplify the implementation, some SIM card vendors (Gemalto, etc.) offer an applet which fully implements steps 1 to 17 (EAP dialog). Also, the security is improved over classical SIM authentication as three RANDs are involved.

8.4 Activating "Circuit Mode" Services (VoIP and SMS)

The terminal has a softphone which need only to be provisioned with the "user name" (the MSISDN of the customer) and the "proxy" (the address of his VoIP server). There is *no need for a password* as VoIP will be used after an EAP-SIM authentication.

8.5 Tunneling the WiFi Internet Traffic into the Normal GRPS Traffic for Volume Charging: Works Only If the Terminal Uses a Tunnel Client

The system would be in fact the full WiLAN access part of a LTE network. It replicates the function of a MME so that whenever a user starts Internet HTTP services (not SIP which is handled directly above), it will perform an UPDATE LOCATION GPRS with the HLR/HSS and obtain the internet APN. It will ask the creation of a PDP Context, a tunnel and all the data traffic will be tunneled to the home GGSN or PDN Gateway (4G).

All of this has been often described in many papers. These are basis functions in LTE networks.

The tunneling principle is illustrated by Figure 8.5; the system would behave as a SGSN or a MME 4G. But in order that it works, *the terminal IP address must be available for the SGSN,* the same for a whole session.

The "in-flight" current systems are not suitable as the IP address is "natted" (see Section 1.4) by the aircraft router and by the ground satellite gateway and is thus *common to all aircrafts and users*. The IP address received by the IP tunneling system cannot be used to know on which GTP tunnel to send the data. For a given session, the *Port Source address changes also with each HTTP GET request*.

The system behaves (LTE case) as a MME. The fast switching of tunneled IP streams is done at the Linux Kernel level for maximum performances in recent GGSN which integrate the IP router.

For a WiFi access network compatible with GPRS tunneling for data services, the terminal must have a tunnel client and establish a tunnel with the GGSN exactly as in the classical GPRS mode. A valid detailed implementation is given in the Section 4.3.2. *But it can work only with terminals which can enable a tunnel to access the internet services* Only in this case, the normal GPRS charging can be done by the GGSN in the HPLMN. Without a tunnel, internet access can be provided but only with a "Local Break Out" at the ground satellite gateway and charged by a RADIUS system. This is exactly the principle behind *the 3GPP to WLAN interworking* for UMTS and LTE [8.10].

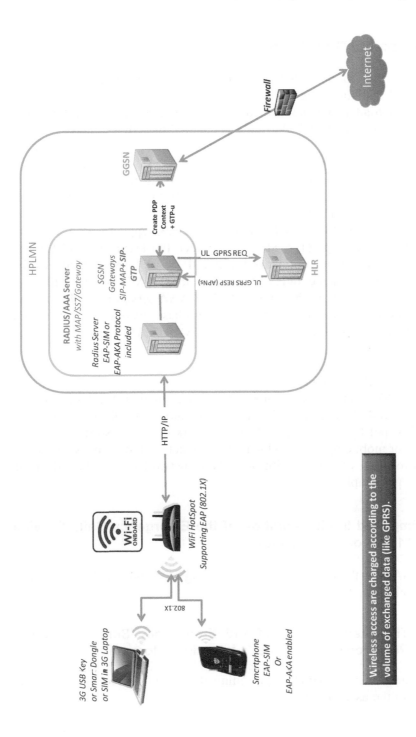

Figure 8.5 Theoretical principle of the tunneling of a WiFi Access through GPRS.

8.6 Support of "WiFi International Roaming"

Here, we refer to not what is *often called "WiFi roaming"* which is "handover" without losing the session from one Access Point to the other, but using abroad a "hotspot" which has made an agreement with the home internet operator, so that WiFi usage is charged directly on the customer's home bill.

This possibility exists in some RADIUS/AAA servers. The Access-Point is always configured to be associated with a given RADIUS/AAA local server (in the visiting country normally). If it receives an "Identity" or "User-Name", at step 1, Figure 8.3, such as

162302001234567@wlan.mnc002.mcc623.3gppnetwork.org

made of the EAP type (1 = SIM, 0 = AKA), the IMSI (Telecel Central Africa) and the "realm"(optional), the RADIUS/AAA server analyzes the IMSI and using the same logic as for international MMS routing, will relay the RADIUS message to the home RADIUS/AAA server, as well as all the request and responses of the authentication phase dialogue which is then between the   terminal (in France) and the RADIUS/AAA server in Central Africa.

Who can use this system?
The MAP/SS7/Gateway has some VLR functions with the HLR. So the system can only be used by a full mobile operator or a full MVNO with its own HLR, not by a WiFi or WiMAX network operator. So the system is suitable for mobile operators which opiates a national WiFi network or has a WiFi roaming capability with other WiFi networks as explained in the previous paragraph.

8.7 Simplified Implementation of the Emergency Calls Service (911(US) or 112(Europe)

The PSAPs (Public Service Answering Points), police, various emergencies, need to locate callers.

8.7.1 The Standard 911 of 112 Emergency Service and the Requirement for Caller's Location Information

The security services (PSAP) providing these services need to have the location of the user for three purposes:

- directing the call to a center using the same language (if a US subscriber calls 112 from France, route him to an English speaking person),
- avoid malicious calls,
- have a location estimate sufficiently accurate to alert the closest appropriate assistance (fire station, police, etc.).

Providing the location information is part of the MNOs obligations in most countries.

The emergency caller will generate a call to the PSTN most often using ISUP. Usable Location information may be contained in the parameter 'Location Number' of the ISUP calls received by the emergency services and provides a simple way to get an approximate location. The parameter content is standardized in only a few countries. The lack of international and even national standardization of the location number is shown by Table 8.1 and Table 8.2.

Table 8.1 Examples of Location Number in ISUP messages (France)

Mobile Operator	Example of « Location Number » parameter in the call setup ISUP message	Interpretation
Orange	61 75015 00	61(Orange) ZIP code of BTS or NodeB
Bouygues	63 92000 00	63(Bouygues) Generic ZIP code of the visited VLR
SFR	62 75015 00 57804	62 (SFR) ZIP code + Cell ID
Free Mobile		

Table 8.2 Examples of Location Number in ISUP messages (Sweden)

Mobile Operator	Example of « Location Number » parameter in the call setup ISUP message	Interpretation
Telia Sonera	46705000111	GT of the visited VLR
Tele2 Comviq	46 707 56201 73281	CC+MGT + LAC + Cell Id

For the examples, SFR provides emergency services with the most accurate information (the PSAP has a map of the cell numbers) and Bouygues with less precision. In some countries, MNOs have their own SMLC (Serving Mobile Location Center) providing an accurate location (better than 150m in the US as specified by the FCC). The PSAPs have an interface for accessing these SMLCs. Their systems can use the Calling

Party Number (even if the display is "hidden", the information is always there) and can then ask a more accurate information based on the method that they offer, at least the LAC-Cell Id and eventually the GPS position contained in the mobile or calculated by A - GPS through their SMLC.

8.7.2   Location Information Provided for a VoIP/WiFi Call

8.7.2.1   Case of GSM Access Deactivated

The only information known by the central server when users register on an IP access is:

- IP address of the Access Point (except when it is "natted" as the case on in-flight WiFi, the IP is that of the ground satellite gateway, whatever the aircraft)
- The IP of the handset allocated by the Access Provider (it is one of their local addresses).

Some have used the IP address to obtain a crude location (the country) by interrogating the reverse registers 'who is'. From the IP address it gives the Access Provider and the country. This is known *not to be accurate because the access may use "proxy" which changes the address*, but is *applicable for commercial purposes* such as the websites setting automatically the language depending on the customer's country. The first patent [8.9] in 2001 was for access authorization to certain commercial web services based on the Access Provider name.

More recently (see Figure 8.6), it has been used in certain VoIP services, the server may send a welcome message to the softphone giving the Access Provider name and the country. If the service price depends on the country it may be also useful to display the rate. It is easy when the ISUP call is made to the emergency service in the Home Country through the PSTN, to create a Location Number with the Visited Country Code.

An example of a forced location number in the ISUP call is given in Table 8.3. It is easy when the ISUP call is made to the emergency service

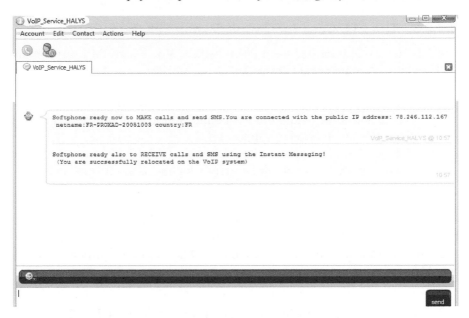

Figure 8.6 Example of Access Provider display on a softphone..

Table 8.3 Location Number in an ISUP call from VoIP

Access Provider Name	Example of « Location Number » parameter in the call setup ISUP message	Interpretation
Singtel	46	Visited Country Code

in the Home Country through the PSTN, to create a Location Number with the Visited Country Code.

8.7.2.2   Case of GSM (Circuit Domain) Still Active

However, there are VoIP service implementation cases, when *VoIP is only used for outgoing calls and outgoing SMS*. To receive calls and SMS, outbound subscribers keep their GSM active but use VoIP for outgoing calls. The implementation is then very cheap (just a very standard VoIP server, without any need for the MSC/VLR part) for cheap subscription customers. In this case as there is also an active GSM connection, the traditional crude location methods may be used (Cell-ID) when requested by the PSAP when an emergency call is made through WiFi.

8.7.2.3   The LTE Case for Circuit Switch Fall Back(CSFB)

Until 2010, two approaches were discussed for Voice and SMS services in LTE (which is a pure packet network):

- VoLTE is obviously simpler: it uses SIP over IP connections for voice and SMS, with no other connection to a mobile network: the handsets just needed to support LTE
- CSFB: the LTE handset also includes a GSM 3G handset which is connected to a 3G Circuit Domain which is used for calls and SMS. The usual 3G location information may be obtained by the PSAP.

Because they were bound by their obligations, including the provision of the emergency call services with location information, many operators including European ones, had to choose CSFB, even though pure VoLTE would have been much simpler.

8.8 Setting up Remotely the Terminal for WiFi with EAP-SIM Network Using OTA

As in Section 1.6, it is possible with OTA-GPRS to setup EAP-SIM in a terminal. Here is an example for an Orange France subscriber using the OMA standard and the usual password method (here, the "WEPKEY" A13467F00D). This uses four SMS sent to port 2948. For a manual setting, the user sees in his telephone a list of WiFi networks (their SSIDs) and selects the access method: WEP and WPA for password methods, EAP-SIM or EAP-AKA (no need).

```
OTA GPRS(using the 'OMA standard') received in 4 concatenated SMS
?xml version="1.0"?
!DOCTYPE wap-provisioningdoc PUBLIC "-//WAPFORUM//DTD PROV 1.0//EN"
"http://www.wapforum.org/DTD/prov.dtd"
< wap-provisioningdoc version="1.1" >
 <characteristic type="BOOTSTRAP">
            <parm name="NAME" value="AHLwifi"   />
 </characteristic>
 <characteristic type="NAPDEF" >
            <parm name="NAPID" value="AHLwifi" />
            <parm name="BEARER" value="WLAN" />
            <parm name="NAME" value="AHLwifi" />
            <characteristic type="WLAN">
               <parm name="PRI-SSID" value="MONWIFI" /> /* the SSID of the WiFi access Point
               < par
            m name="NETMODE" value="INFRA"   />
               <parm name="SECMODE" value="WEP"        /> /* 802.1x for EAP-SIM
               < parm name="WEPAUTHMODE" value="OPEN" />
```

```
                <parm name="WEPKEYIND" value="0" />
                <characteristic type="WEPKEY">
                        <parm name="LENGTH" value="64"     />
                        <parm name="INDEX" value="0"        />
                        <parm name="DATA" value="A13467F00D" />  // the password
                </characteristic >
                < parm name="EDIT-SET" value="0"  />
                < parm name="FORW-SET" value="0" />
                <parm name="VIEW-SET" value="0" />
        </characteristic>
</characteristic >
<characteristic type="APPLICATION">
        < parm name="APPID" value="w2" />  /* internet acces type profile
        <parm name="TO-NAPID" value="AHLwifi" />
        <characteristic type="RESOURCE">
        < parm name="NAME" value="orange" />
        < parm name="URI" value="http://www.orange.fr" />  //the home page
        < parm name="STARTPAGE" />
        </characteristic>
</characteristic>
< /wap-provisioningdoc>
```

To provide an EAP-SIM access for WiFi, replace SECMODE by 802.1x and the WEPKEY set of parameters (the password) is not required.

If EAP-SIM needs to be provided to all users in a WiFi aircraft with EAP-SIM, it is relatively simple as part of the welcome message to send automatically the EAP-SIM setting if it is a first time user. If a tunnel client needs to be configured, it can be included in the same setting.

References and Further Readings

[8.1]   A.Henry-Labordère, "Système de Terminaisons d'appels vers des numéros de mobiles par IP sans coopération du réseau de mobiles", Patent FR 07 301 546 3

[8.2]   A.Henry-Labordère, S.Cruaux, G.Deviercy, B.Mathian, T.Braconnier, "Système de communication téléphonique par VoIP", Patent 1257853

[8.3]   P.Urien, "La Sécurité des réseaux sans fil 802.11", Les Cahiers du Numérique, Lavoisier ed., 2003/3 Vol 4, pp 167-184 (a very good introduction to EAP-SIM with call flows)

[8.4]   RFC 3579, RADIUS(Remote Authentication and Dial In User Service) Support for Extensible Authentication Protocol(EAP), Sept 2003.

[8.5]   RFC 2869, RADIUS Extensions", June 2000

[8.6]   RFC 4186, Extensible Authentication Protocol Method for Mobile Communications (GSM) Subscriber Identity Modules(EAP-SIM), Jan 2006 (gives all the details and examples to implement EAP-SIM).

[8.7]   RFC 2548, Microsoft Vendor-specific RADIUS Attributes, March 1999 (explain the details of the mandatory computation of the Unicast Keys used to secure any WiFi radio acces)

[8.8]   RFC 5448, Extensible Authentication Protocol Method for 3rd Generation Authentication and Key Agreement (EAP-AKA), May 2009 (gives all the details and examples to implement EAP-AKA for USIM).

[8.9]   A. Lardenois, Patent FR 2806178 (2001), « Méthode d'identification d'un appel internet vers un serveur http, pour permettre des autorisations ou interdictions d'accès »

9

GSM<>Satellite handset seamless number continuity

The gravity law is tough, but it is the law

> – *A policeman*

9.1 User Benefit from the GSM<->Satellite Number Continuity Implementation

A customer owns two mobile phones: one GSM mobile and one satellite phone. His contacts know only his GSM number and he will receive few calls to his GSM numbers as no one knows he is using a satellite phone.

He is currently in an uncovered GSM area, ship, or open country but his satellite phone is activated after he has turned it on. With the Number Continuity implemented for his satellite phone, he will receive all calls and SMS to his satellite phone even though they were sent to his GSM number. The only thing he has done is turning on the satellite phone. There is no manual call forwarding (it would not forward the SMS) and if he loses GSM coverage he would not be able to use the call forwarding function. If he is again under GSM coverage, he could reactivate it, and again all calls and SMS will be received by the GSM handset.

The Number Continuity Solution allows this transparent service for GSM and Satellite phone customers. It is implemented by satellite operators, or by Roaming Hub operators in cooperation with them and with agreements with the involved GSM operators.

9.2 Satellite Coverage and Position Station Keeping of GEO and LEO

9.2.1 Telecom Coverage

For GEO and LEO, the orbit plane is supposed to be the equator (which makes an angle of 23° 27' (the tilt if it is the earth axis) with the earth ecliptic plane, where all the other planets are located mostly (Pluto has an inclined orbit). For a LEO satellite, the orbit is strongly inclined at the equator.

From Newton's gravitation law:

$\gamma_g = K/R^2$ for the terrestrial acceleration where R is the geostationary satellite *distance* from the earth center. The gravitation field decreases as the square of the distance to the earth center.

$\gamma_c = \omega^2 \times R$ is the centrifugal acceleration applied to the satellite due to its circular assumed orbit around the center of an earth inertial coordinate system.

The earth (same angular speed exactly as satellite) has a rotation time in one sidereal day $D = 23h$ 56m 4s (not exactly the legal 24 hours), which gives the common angular speed ω of the earth and of the geostationary satellite:

$\omega = 2\Pi/ D = 0.000072921173$ rad / sec.

K is such that $\gamma_g = 9.822$ m/sec^2 (the *standard gravity*) at $R_o = 6\ 371$ km the official geodetic radius of the "geoïd surface", so:

$K = 9.822 \times (R_o)^2 = 3.986714 \times 10^{14}$ (in m^3/sec^2).

Setting the gravitation equal to the centrifugal acceleration $\gamma_g = \gamma_c$ for a circular orbit, with the above values of K and ω, yields:

$R = (K / \omega^2)^{1/3} =$ 42166 km (distance from the earth center)

which gives an *altitude* of $R - R_o = 35795$ km above the standard ground level if the orbit was circular, commonly approximated by 36000 km and it is independent of the mass of the satellite.

If the orbit was circular and the inclination of the orbit is equal to $0°$ with the equator, the satellite would look still from a ground telecom station, which is simple for tracking.

The satellite covers the earth within a cone which ½ angle is:

Arc cos $(R_0 / R) = 81°$ latitude

This is a higher latitude coverage of the north zone that most LEO systems with inclined orbits.

9.2.2 West–East Drift (Deviation from a Theoretical Circular Orbit)

When the satellite is launched, the orbit is not exactly circular (this is a delicate and fuel consuming launch phase from an elliptical "transfer orbit") even if we assume that the revolution time is exactly one sidereal

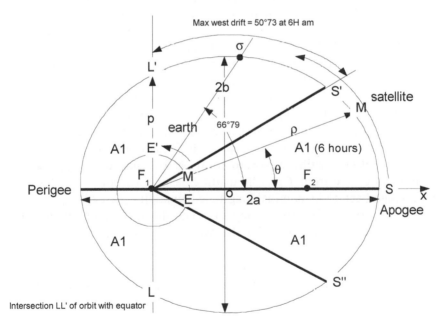

Example of defective geostationnary orbit: excentricity too large = -0.507

Figure 9.1 Geostationary orbit with eccentricity e not 0.

day. The satellite has an elliptical orbit (Kepler's ellipse law) with a small excentricity e ($e = 0$ for a circle, and $e < 0$ for an ellipse). Its equation in polar coordinates is

$\rho = P / (1 + e \cos \theta)$

P is the distance from the focus F_1 (the center of the earth) to the ellipse. To test real numerical examples, here are the classical relations between the parameters of an ellipse in polar coordinates and in Cartesian coordinates:

$P = a (1 - e^2)$, easy to find, by computing ρ for $\theta = 0$ (apogee) and $\theta = \pi$ (perigee)

$P = b^2 / a$

$c = a\,e$ (½ distance between the two focus F_1 and F_2)

A (the area inside the ellipse) $= \pi$ ab

The perigee (lowest altitude) is P and the apogee (highest) is A.

Kepler's law of equal areas says that an equal area of the ellipse is covered in a given time interval. Assume that at 0h, the satellite is in S vertical of the earth E (the nominal satellite longitude). 6h later, the satellite is in S' (the area $A_1 = AS'F_1$ is equal to the area $F_1S'P$ as P is the satellite position at 12h). At 0h, 12h and 24h again, the satellite is at its nominal longitude. But in the AS' part of the day, the satellite is revolving slower that the earth and appears to drift westward (the earth turns west to east). When the satellite is in S' the drift difference is the angle EF_1S'. In the $S'P$ part, it is faster and appears to drift eastward. An earth observer sees the satellite drifting westward form 0h to T and coming back to its nominal longitude from T to 12h, drifting eastward from 12h to $(24h - T)$ and coming back to its nominal longitude from 24h $- T$ to 24h.

If the orbit is far from being circular, the $E\text{-}W$ telecommunication coverage may be defective during the two daily periods of maximum drift.

Angular position $\theta(t)$ of the satellite

The total area A of the ellipse can be computed simply as above knowing P and e. The differential of A for a small angle $d\theta$, is

$dA = \frac{1}{2} P^2 \, d\theta$ (area of a small triangle with height ρ and basis $\rho d\theta$)   [9.1]

To get $\theta(t)$, we use Kepler's equal area law.

Assuming to simplify that the angular position of the satellite is $S = 0$ at 0h, we have

$$S \frac{1}{2} \left(\frac{p^2}{(1 + e \cos \theta)^2} \right) d\theta = (A/D) \ t \text{ the area of the ellipse from 0 to } t \text{ time) 0, } t$$

Kepler's law: Area between time 0 and $T = \int_0^T \frac{p^2}{(1 + e \cos \theta)^2} d\theta = kT$ [9.2]

Taking $T = D$ (full revolution in one day 23h 56m 4s), the total ellipse area being A,

$$A = kD \rightarrow k = \frac{A}{D} \tag{9.3}$$

This is the integral equation relating to time T and the angular position $\theta(T)$.

It is complicated to integrate by hand but this is a "definite integral" (using elementary functions), the online formal integration tool [9.3] gives the result using real valued functions only for $|e| > 1$ so the trick is to set the eccentricity $e' = 1/e$ in [9.2] and do instead a formal integration of:

$$g(\theta) = \int \frac{1}{\left(1 + \left(\frac{1}{e} \right) \cos \theta \right)^2} d\theta = \tag{9.4}$$

which gives an expression of real valued functions ($e^2 - 1$ is > 0):

$$e^2 \left(\frac{2e \tan^{-1} \left(\frac{(e-1) \tan\left(\frac{\theta}{2} \right)}{\sqrt{e^2 - 1}} \right)}{(e^2 - 1)^{\frac{3}{2}}} - \frac{\sin(\theta)}{(e^2 - 1) \cos(\theta) + e} \right) \tag{9.5}$$

Using Kepler's area law [9.2], the value of k in [9.3] and the expression of $g(\theta)$ [9.5] yields:

$$g(\theta(T)) = \frac{A}{DP^2} T \tag{9.6}$$

This gives

$$\theta(T) = g^{-1} \left(\frac{A}{DP^2} T \right), \text{ (but } g^{-1} \text{ can only be computed numerically)} \tag{9.7}$$

C program to compute the time T to have the angle θ

```
/*-------------------------------------------------------------------------------  */
/* Computes the time in hours for the satellite to go from 0 to theta   */
/* on an elliptic orbit following Kepler's law                          */
/* AHL 22/11/2012                                                       */
/* Method: use the primitive obtained by formal integration, see [9.5] */
/* then applies Kepler's law to obtain the time                        */
/* ENTRY ; theta : angle in radians                                    */
*/
/*              eprime: eccentricity of ellipse -1 < e < 0(circle)      */
/*              P ellipse parameter in km                               */
/*              AS : 1/2 area of ellipse reached at day/2, in km2       */
/* constant : D (in 1/2 days, with day) is 23H 56min 4 sec              */
/* RESULT: TIME in hours                                                */
/*-------------------------------------------------------------------------------  */
double Temps(double theta, double eprime, double P , double AS)
{
  double TIME, Prim;
  double e, e2moins1;

  e = 1/eprime; //inverse of eccentricity is then < -1 for an ellipse
  e2moins1 = e*e - 1; // which is > 0 !! used to simplify the expression of
the primitive
// result of formal integration to obtain the primitive
  Prim = (e*e)*(
     (2*e/(e2moins1*sqrt(e2moins1)))*atan((e-1)*tan(theta/2)
/sqrt(e2moins1))
        - sin(theta)/(e2moins1 * (cos(theta) + e))
        ); //thanks to Pascal Adjamagbo and Jean-Yves Charbonnel
(Institut de Mathématiques de Jussieu(IMJ) nov. 2012

  TIME = D * (0.5 * P*P * Prim) / AS; //Kepler's law :time in hours from 0
to theta as a proportion of the 1/2 ellipse area AS
  return(TIME);
}
```

Computation of angle of radial speed synchronism

The radial speed of the satellite and of the earth is the same for an angular position θ_s such as

$$\frac{P^2}{(1+e\cos\theta_s)^2} = \frac{A}{\pi}$$

that is

$$\theta_s = \text{Arc cos}\left(\frac{P\sqrt{\frac{\pi}{A}}-1}{e}\right)$$

which we represent as σ in Figure 9.1.

Computation of the maximum westward and eastward daily drift.
The drift is the difference in longitude between the normal longitude of the geostationary satellite and the real longitude θ_s = 66° 79 with the above numerical values. At this angular position, the west drift is maximum because the satellite becomes faster eastward than the earth.   The numerical integration gives t_s = 8H27 min. We can compute the earth angular position and the drift.

Normally, the tolerance given by the space organizations for geostationary station is about 1° which means that if the launch does not succeed in setting an almost circular orbit, the satellite is not usable.

Long term West or East drift
The drift is due to the earth potential not being symmetrical (the equator is slightly elliptical), and has a tendency to "pull" the latitude toward two stable "tesseral points"(from the harmonic function theory of Simon

Table 9.1 West–East and North–East drift depending on the orbit's eccentricity

Excentricity	S' = maximum westward drift at 6H am	Duration of North drift = g(π/2 - 0)	Duration of South drift = g(π–π/2)	Angular position of Synchronism θ_s
– 0.507	60° 08	9H 42min	2H 18min	66° 79
...	...	...	...	
– 0.017	1° 95	6H 8min	5H 52min	89° 27
– 0.012	1° 38	6H 5min	5H 55min	89° 48
– 0.007	0° 80	6H 3min	5H 57min	89° 69
– 0.002	0° 23	6H 1min	5H 59min	89° 91
0 (circle)	0°	6H	6H	All

Laplace) at 75 °E and at 104 °W, with two unstable "tesseral points" at 165°E, and at 14°W. Regular fuel consuming corrections must be done from the longitude station keeping.

9.2.3 North–South Drift (Orbit Not in the Equatorial Terrestrial Plan)

The launch may be such that the satellite orbit makes a small angle γ with the equator plane. In Figure 9.2, the intersection of the orbit with the equator is LL' (we assumed for simplicity that the apogee of the orbit was at the nominal longitude). The maximum north latitude deviation is δ at 0H and south at 12h. It remains North from A (apogee) to L' and south from L' to P (perigee). The time during which there is a North–South drift is also

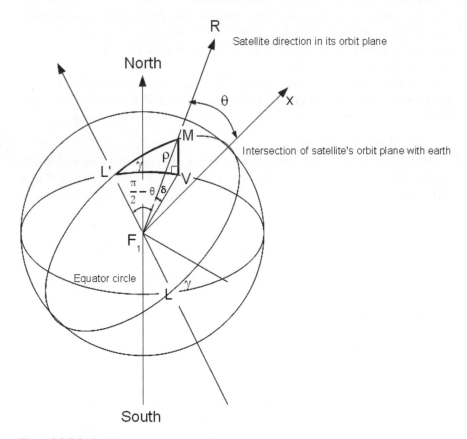

Figure 9.2 Spherical triangles used in the orbit computation.

computed numerically in Table 9.1, for the North drift it is the time when $\theta = {}^{\pi}/_2$ starting from the apogee P at 0h. For $e = -0.507$ the satellite has a 0° latitude at 9h 42 min and 14h 18 min.

North–South deviation during the day
The North–South drift is the "declination" δ of the satellite (corresponding to VM) for a given angle θ. The spherical triangle L'VM (with a rectangular angle V such as sin $V=1$) has the well-known spherical trigonometric relation:

$$\sin \gamma / \sin \delta = \sin V / \sin\left(\frac{\pi}{2} - \theta\right)$$
$$\sin \delta = \sin \gamma \cos \theta$$

("declination" δ of satellite as a function of its angular position θ)   [9.6]

Long term North–South drift
The combined effect of the sun gravitation mainly, of the moon and (the earth potential not being spherical being the least important) tends to "pull" the orbit plane to align with the ecliptic plane earth–moon–sun at a speed of about 1°/year. The fuel budget to maintain the latitude is much more than the longitude station keeping (about 50 m/sec against 2m/sec) speed impulses.

9.2.4 Daily Ground Trace of the "Geostationary" Satellite (Ephemeris)

Not counting the long term drifts, the ground trace is a combination of the 24h period longitude and latitude drifts. In the simplified example we assume the satellite is exactly at his nominal longitude at the apogee (and perigee). We have, using [9.7]

$$\text{Longitude drift} = (2\pi \ T \ / \ D) - g^{-1}\left(\frac{A}{DP^2}T\right) \text{ (difference at time } T,$$
between the earth angular position and satellite's which is $\theta(T)$)

$$\text{Latitude drift} = \delta(T) = \text{arc sin } (\sin \gamma \cos \theta(T))$$

that is using also [9.7]:

$$\text{Latitude drift} = \delta(T) = \text{arc sin}\left(\sin \gamma \cos g^{-1}\left(\frac{A}{DP^2}T\right)\right)$$

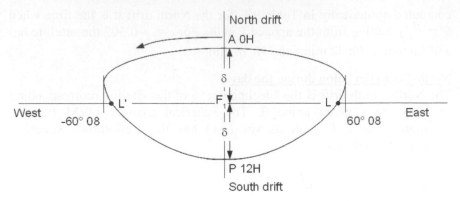

Figure 9.3 Ground trace of geostationary satellite.

If one wanted to use a satellite to make a traditional computation of its own position with a sextant and accurate watch, the curve gives the ephemeris coordinates α and δ of the satellite.

We gave an elementary but exact two-body model (earth and satellite) of the orbit computation. In [9.4], Vol I, there is the more accurate model using the "*n*+1 body" perturbation theory of Lagrange which is used for long-term planet orbit computations and operational satellite ephemeris. When the orbit is close to the simple two-body model, there is a system of 3*n* second order non-linear differential equations, which is numerically computed (François-Xavier Lagrange who invented the method (1736–1813), must have had a hard time in solving properly (he was the first) the three-body problem earth–sun–moon).

9.2.5   Handover in Satellite Calls

Handover is the procedure which allows a call to proceed without interruption when the satellite handset coverage conditions change due to the serving satellite's quick change of position, movement of the handset, or atmospheric conditions.

9.2.5.1   Geostationary   Systems   (Inmarsat   (Isatphone), Thuraya)

Each of the satellites may be considered equivalent to a BSC (RNC). Each of the MSC-VLR corresponds to a single ground station with a single BSC. There is a handover when the user moves further East or West than the coverage of the current satellite. The procedure is exactly the standard "inter-MSC handover" [9.2]. The MSC of the new ground station *A* sends a MAP_PERFORM_HANDOVER to the previous MSC *B* giving the next

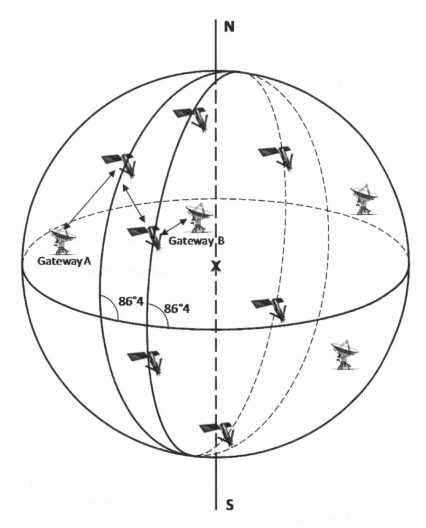

Figure 9.4 Example of a satellite constellation.

cell where the call must be transferred. It is exactly like GSM and described below for the LEO Inter-gateway handover case.

9.2.5.2 LEO Systems (Iridium Example)

What is called a "beam" of these systems has a given frequency and covers a "spot" (earth footprint) for the subscribers. Each antenna (e.g. Iridium case) has 16 beams, so a satellite has a total of 48 beams. Each spot, which may overlap, may be covered by several beams (frequencies).

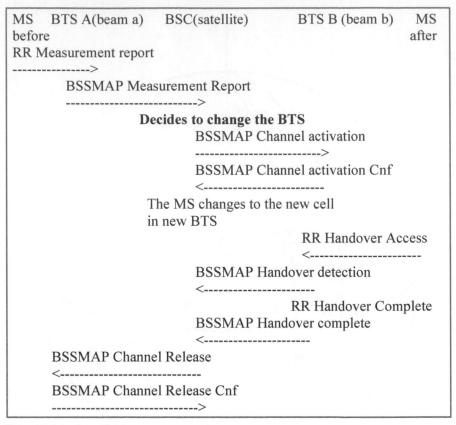

MS   BTS A(beam a)   BSC(satellite)   BTS B (beam b)   MS
before   after
RR Measurement report
---------------->
          BSSMAP Measurement Report
          --------------------------->
                    Decides to change the BTS
                              BSSMAP Channel activation
                              --------------------------->
                              BSSMAP Channel activation Cnf
                              <--------------------------
                    The MS changes to the new cell
                    in new BTS
                                        RR Handover Access
                                        <-----------------------
                              BSSMAP Handover detection
                              <-----------------------
                                        RR Handover Complete
                              BSSMAP Handover complete
                              <-----------------------
          BSSMAP Channel Release
          <-----------------------------
          BSSMAP Channel Release Cnf
          ------------------------------->

Figure 9.5 Inter-beam Handover.

The satellites which have quickly varying positions are connected for control and for relay of communications to a network of XX ground stations, a "satellite gateway", which also are connected to the fixed lines PSTN (calls out and in) and together (inter-MSC handover) by SS7 signaling links through the satellite constellation.

We will explain the LEO handover procedures based on the readers' assumed knowledge of the GSM handover procedures, with a little refreshing. For this comparison: "satellite gateway" = MSC-VLR, each satellite = BSC, each "beam" = cell is covered by a BTS.

9.2.5.2.1 Intra-Beam Handover, Equivalent to GSM "Intra-Number Handover"

As the quality of the signal degrades, the handset may monitor that another frequency is better ("intra-beam handover" at the initiative of the handset) or the satellite wants the handset to use another frequency because it may interfere with another satellite's beam. In this case, the satellite asks the handset to change the frequency.

9.2.5.2.2 Inter-Beam Handover [9.1], Equivalent to GSM "Inter-number Handover [9.2]

The handset decides to switch to another beam of the same satellite because the signal quality is better. In GSM, this is equivalent to a handset going to another cell. It sends a request to the BSC and the BSC manages the change including to a cell it has selected.

9.2.5.2.3 Inter-Satellite Handover [term in 9.1], Equivalent to GSM "Intra-MSC Handover"[term in 9.2]

The serving satellite is moving and the spot where the handset is making a call may not be covered well any more. The serving gateway A which is

```
MS        BSC A(satellite A)      MSC-VLR (ground gateway)       BSC B (satellite B)
  MS
before                                                                         after
           BSSMAP Handover required
           --------------------------------------->
                                    BSSMAP Handover request
                                    ------------------------------------->
                                    BSSMAP Handover request Cnf
                                    <-----------------------------------
           BSSMAP Handover command
              <-------------------------------------------
RR Handover command
<------------
                                    The MS switches to a new cell and frequency
                                                        RR Handover access
                                                        <----------------------
                                    BSSMAP Handover detection
                                       <----------------------------------
                                    The MSC-VLR switches the voice circuit
                                                        RR Handover complete
                                                        <----------------------
                                    BSSMAP Handover complete
                                       <----------------------------------
           BSSMAP clear command
              <-------------------------------------------
           BSSMAP clear complete
           ------------------------------------------->
```

Figure 9.6 Inter-satellite Handover.

```
MS          BSC  A(satellite A)       MSC A       MSC B     BSC B (satellite B)
MS
before                                                                      after
            BSSMAP Handover required
            ------------------------------->
                                   MAP Perform Handover
                                   -------------------->
                                                    BSSMAP Handover request
                                                    ------------------>
                                                    BSSMAP Handover request Cnf
                                                    <---------------
                                   MAP Perform Handover Cnf
                                   <--------------------
                                   MSC A establish
                                   a voice circuit to MSC B
            BSSMAP Handover command
            <------------------------------
RR Handover command
<------------

                                   The MS switches to a new cell
                                    and frequency
                                                            RR Handover access
                                                            <----------------------
                                                    BSSMAP Handover detection
                                                    <--------------------
                                   MAP Process_Access_Signaling
                                     <--------------------

                                                            RR Handover complete
                                                            <----------------------
                                                    BSSMAP Handover complete
                                                    <--------------------
                                   MAP Send_End_Signal
                                   <---------------------
                                   MSC B confirms the voice circuit
                                   creation to MSC A
            BSSMAP clear command
            <----------------------------
            BSSMAP clear complete
            ---------------------------->
                                   The voice calls continues with
                                   A being the "anchor gateway" and
                                   B the "serving gateway"
```

Figure 9.7 Inter-gateway Handover.

used for the call asks a new satellite under its coverage to pursue the call with the handset. The connection of the handset with the previous satellite is released. The gateway acts as a MSC-VLR in "the inter-MSC handover" case (same MSC, new BSC) which is recalled Figure 9.6 below (it is not standard BSSAP which is used for satellites of course).

The radio equivalent of the well-known SCCP is called the protocol RR. As you can see, there is no use of the MAP protocol as only one MSC-VLR is involved and MAP is used only for Handover between MSC-VLR as in the inter-MSC handover below.

9.2.5.2.4 Inter-Gateway Handover, Equivalent to GSM "Inter-MSC Handover[9.2]

Due to the movement of the handset, it is not well covered any more by a satellite covered by the current ground gateway ground gateway where the call was established ("anchor gateway"). This case exists for satellite but this is not described in [9.1]. When a handset finds a better coverage with another satellite, it will start the same as above giving the identity of the new satellite (new LAC. Cell Id in GSM) to his current service gateway A. In the GSM, it will be as Figure 9.7:

The call is "tromboned" using the voice pass through the satellites as illustrated in Figure 9.4.

9.3 Satellite Operators Concerned by the "Number Continuity Service

They provide special handsets with a direct radio connection to satellites. In most cases they also operate a satellite network, either geostationary (GEO) with an equatorial orbit or a constellation of Low Earth Orbit satellites with inclined orbits. Below the explain the basis of satellite dynamics and also the principle of telecom space networks with a constellation of LEO spacecrafts when the handover is required to maintain the connections (Iridium, Globalstar)

9.3.1   Some satellite handsets and core networks

They are easily recognizable by their voluminous foldable antenna. To be practical, it is easy to see that the handsets below have big antennas, they do not look as compact GSM.

The SIM card terminals accept GSM cards from networks if there is roaming agreement. The user is invoiced by his own network. Only the Thuraya handset is usable on fixed GSM network.

- with GSM or CDMA compatibility (roaming possibility with terrestrial networks):
    - Globalstar, (LEO constellation), (GSM with SIM card or CDMA, only the CDMA handset is commercialized any more) (commercialisation and national gateways (eg. TESAM (GSM) France closed in 2001, see list in above tables
    - Thuraya (GSM)
    - Innmarsat (C et M terminals), and Isatphone handsets with SIM card.
    - Thuraya (GSM) with SIM cards.
    - AcES (GSM) with SIM cards (closed in 2004, service taken by Inmarsat with new satellites and ground infrastructure under the Isatphone brand.

GLOBALSTAR Qualcomm
GSP-1700 CDMA
no SIM card ;
no CDMA radio compatibility

GLOBALSTAR Telit
SAT550 GSM (with SIM card)
no GSM radio compatibility

INMARSAT Isatphone
Pro GSM with SIM card
no GSM radio compatibility

IRIDIUM 9555
Satellite GSM (with SIM card)
no GSM radio compatibility

THURAYA Hughes
7101 GSM (with SIM card)
has a GSM radio interface

Picture 9.1 Various current satellite phones, some with large antennas

The various satellite networks are compared, in Table 9.2 below, note in particular the differences in orbit types. While Table 9.3 gives details on the numbers of satellites and the satellite vendors.

Table 9.2 List of satellite operators

Name	Type of orbits	Number of satellites	Core network Technology and vendor	Ground Gateways (with GMSC)	Services provided
Globalstar	LEO	48	GSM (Alcatel) and IS-41(CDMA) DSC Communication Corporation	France (Aussaguel), Russia, Chili, Turkey, USA (Texas)	Voice SMS no Data
Iridium	LEO	66	GSM (Ericsson)	Tempe, Arizona Wahiawa, Hawaii — owned by DISA Avezzano, Italy Pune (India), Beijing (People's Republic of China), Moscow (Russia), Nagano (Japan), Seoul (South Korea), Taipei (Taiwan), Jeddah (Saudia Arabia) Rio de Janeiro (Brazil)	Voice SMS Data
Thuraya	GEO	2	GSM (Hughes)	Emirates	Voice SMS no Data
Inmarsat (Isatphone)	GEO	4	GSM (Ericsson 3G HLR-MSC-SGSN-GGSN+ Lockeed Martin)	Bochum Hawaii + 1	Voice SMS Data
AcES (closed)	GEO	1	GSM	Indonesia, India, Taiwan	Closed

Table 9.3 Table of satellite vendor's for the satellite telephone service

Networks	Satellites	Designer/vendor of mobile
Globalstar (roaming entre les Gateways Globalstar): on peut avoir un numéro +336400x et s'en servir en Australie	52 Loral satellites (originally) 32 Thalès Alenia (new generation), inclined 52° in 8 orbit planes.	Qualcomm (based on a CDMA handset, no Globalstar SIM card, the MSISDN Globalstar is wriiten in the handset. Telit (GSM)
Iridium	LEO (near polar, inclined 86°4), only system covering polar regions. 66 satellites in 11 orbit planes. Thalès Alenia(nouvelle génération)	
Thuraya	Boeing (2 satellites)	ASCOM and Hughes
Inmarsat	3 Astrium EADS(4 I-4 et 7 I-2 ou I-3)	Elcoteq(Estonia)
AcES (closed 2011) had roaming with : Hong Kong CSL Hutchison Telecom (HK) Bharti Hexacom Ltd (AIRTEL) Excelcomindo(Indonesia) PT Indonesian (INDOSAT) PT Telekomunikasi Selular (TELKOMSEL) Safaricom (Kenya) DiGi Telecommunications (Philippines) SingTel Mobile Singapore Dialog Axiata (Sri Lanka) Swisscom (Switzerland)	« Garuda » was the satellite name	Ericsson

The GSM handsets have a SIM card and are dual mode GSM and satellite radio transmission. Technically, Thuraya does it. They can visit a terrestrial GSM network if they have the roaming agreement.

- without GSM or CDMA compatibility(cannot use terrestrial networks)
  - Iridium (constellation)

9.4 Operators Not Concerned by the Need for Number Continuity: Air, Sea (Maritime) and GSM "Bubble Service" and Satellite Operators without Voice Services

On the contrary, ONAIR, AeroMobile, Meagafon which provide the GSM service (with "femtocells" in the aircraft or ships are not concerned by "number continuity" unless they use WiFi. The ships are equipped with classical BTS and leaking lines are used as antennas. There is a satellite radio link (Inmarsat mostly) with the ground segment. This allows using the standard GSM handsets. Astrium is a "GSM bubble service" provider; they install ground BTS with satellite links (their Astrium or Inmarsat) to their core GSM network (HLR, MSC and IN) .

On the GSMa site (2012), in the Air category there are three operators — ONAIR, Aeromobile and MegaFon.

In the Sea category there are seven operators — Maritime Partners, ATT(Wireless Maritime), Siminn (On Wave), ONAIR(OnMaritime), Seanet, Smart Coms(Blue Ocean), Telecom Italia.

Also, even though Intelsat is a major satellite operator, they are not concerned (no handset) neither the access providers such as Satcom.

However the air, sea and "bubble" operators are concerned with the alternative use of WiFi by smartphones to provide the same service as GSM. The pure data services (no voice) operators are given by Table 9.4.

Table 9.4 Pure data services satellite operators

Operator	Type of satellites	Type of service
Intelsat	66 GEO	
Orbcomm	29 LEO (775km). Small satellites (50 – 120 kg) launched mid 1990s.	Small amount of data (messages) and Automatic Identification System (simple devices). All boats in the Vendée Globe race are equipped with an IAS which helps to avoid collisions.

9.5   Coverage and Details of the Various Satellite Operators

9.5.1   Globalstar

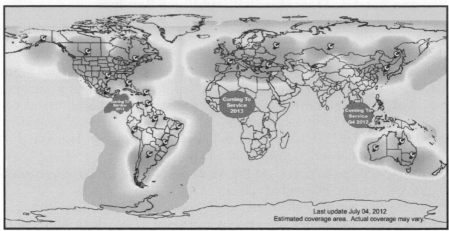

Source: http://www.globalsatellitecommunications.com/globalstar/coverage_map.html

Picture 9.2 Globalstar coverage.

9.5.2   IRIDIUM

➢ Iridium SSC, Iridium communications service was launched on November 1, 1998. Motorola provided the technology and major financial backing.
➢ Chapter 11 bankruptcy nine months later, on August 13, 1999.
➢ Service was restarted in 2001 by the newly founded Iridium Satellite LLC, which was owned by a group of private investors.

The Iridium constellation is the largest in the world, with 66 low earth orbiting (LEO) satellites operating as a fully meshed network. Iridium has flexible billing and flat rates for calls from anywhere to anywhere on earth.

GSM-like offer

Removable Subscriber Identity Modules (SIMs) are used in Iridium phones, much like those used for GSM. Prepaid SIM cards are usually green while post-paid cards are red.

Source: http://www.iridium.com/support/library/CoverageMaps.aspx

Picture 9.3 Iridium beam coverage

Iridium operates at only 2.2 to 3.8 kbit/s, which requires very aggressive voice compression and decompression algorithms. Latency for data connections is around 1800 ms round-trip, using small packets.

There is a web/e-mail to <u>SMS gateway</u> which enables messages to be sent from the Internet or an e-mail account to Iridium handsets for free. There is also a voice mail service.

Tracking transceiver units

Without an extra GNSS receiver tracking is difficult, but not impossible, as the position of a mobile unit can be determined using a Doppler shift calculation from the satellite. These readings however can be inaccurate with errors in the tens of kilometers. Even without using Doppler shifts, a rough indication of a unit's position can be found by checking the location of the spot-beam being used and the mobile unit's timing advance.

The position readings can be extracted from some transceiver units and the 9505A handset using the -MSGEO AT command.

9.5.3   Thuraya

Thuraya is an international mobile satellite services provider, based in Abu Dhabi and Dubai in the United Arab Emirates, and covers mainly the Middle East, Africa, Western Europe, Asia and Australia.

The system allows telecommunications in voice, data and SMS. The services also provide the GmPRS for direct access to the Internet.

Several models are available; they allow the connection by satellite as well as the GSM networks.

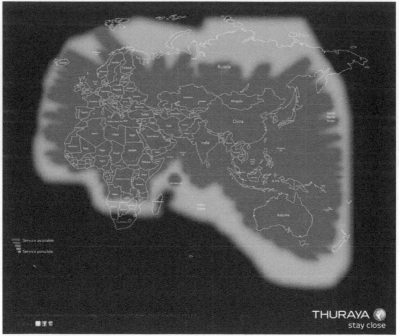

Source: http://www.thuraya.com/coverage-map

Picture 9.4 Thuraya coverage

9.5.4   Inmarsat

INMARSAT coverage foot print
The MAP gives the number of the "spots" which can be used to get a very rough estimate of the handset position with the timing advance.

Systematic Legal Interception
USA, Russia, India and China enforce that the aircraft using GSM/ Inmarsat, when they fly over their territory have all the voice and data communications "tromboned" through their monitoring system before coming back to the ground station concerned. This is automatic as the satellite links transmit permanently the coordinates of the aircraft and the ground station, which has a numerical map, automatically establish and suppress the "tromboning".

The satellites are digital transponders that receive digital signals, reform the pulses, and then retransmit them to ground stations.

Ground stations maintain usage and billing data and function as gateways to the public switched telephone network and the Internet.

Country codes
The permanent telephone country code for calling Inmarsat destinations is 870 SNAC (Single Network Access Code). The 870 number is an automatic locator; we need not know to which satellite the destination Inmarsat terminal is logged-in.

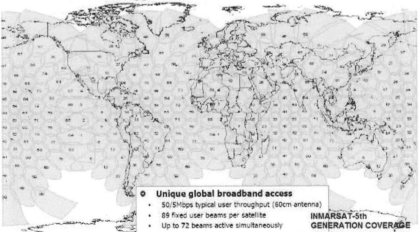

o **Unique global broadband access**
- 50/5Mbps typical user throughput (60cm antenna)
- 89 fixed user beams per satellite
- Up to 72 beams active simultaneously

INMARSAT-5th
GENERATION COVERAGE

Source:http://www.inmarsat.com/cs/groups/inmarsat/documents/document/016329.pdf
http://www.groundcontrol.com/Global_Xpress_Coverage_Map.htm

Picture 9.5 Inmarsat coverage.

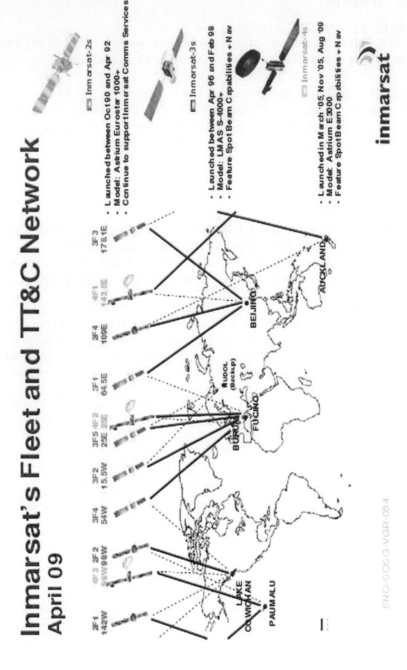

Picture 9.6 Inmarsat tracking stations network.

Land portable

	Wideye™ Sabre™ I' Voice and data, single-user device	EXPLORER® 300 Highly compact, robust device	EXPLORER® 500 High bandwidth, highly portable device
Manufacturer	Addvalue Communications www.wideye.com.sg	Thrane & Thrane www.thrane.com	Thrane & Thrane www.thrane.com
Size	259 x 195mm (1.6kgs)	217 x 168mm (1.4kgs)	217 x 218mm (1.4kgs)
Standard IP	Up to 240/384kbps (send/receive)	Up to 240/384kbps (send/receive)	Up to 448/464kbps (send/receive)
Streaming IP (send & receive)	32, 64kbps	32, 64kbps	32, 64, 128kbps
Voice	Via RJ-11 or Bluetooth handset/headset	Via RJ-11 or Bluetooth handset/headset	Via RJ-11 or Bluetooth handset or 3.1kHz audio/ fax
ISDN	N/A	N/A	64kbps via USB
Other data interfaces	Ethernet, Bluetooth	Ethernet, Bluetooth	Ethernet, USB, Bluetooth
Ingress protection	IP 54	IP 54	IP 54

Picture 9.7 Inmarsat data stations.

Gateways
The fixed part of the ground segment comprises one or more gateways to access the space segment, transport the necessary signaling, control and communications, and support inter/intra system mobility. The location of

gateway depends on operator's preferred selection criteria. Factors considered include proximity to the terrestrial traffic, desired network connectivity and routing arrangements.

Gateways comprise a radio system to support transport over the satellite system and a network switching system to interconnect to the terrestrial network. The satellite mobile phone uses the IP protocol.

The gateway is interfaced to various terrestrial networks such as Public Switched Telephone Network (PSTN), Public Land Mobile Network (PLMN) such as GPRS or UMTS, Internet, private network, etc. The service providers, responsible for end-to-end service provision, access the satellite network through the appropriate interfaces.

BGAN (Global voice and broadband data) use the Inmarsat-4 satellites. They are used from laptop-sized terminals.

I-4 traffic is transported mainly in the form of IP (Internet Protocol) data packet, thereby increasing the capacity of the Inmarsat network to provide advanced digital mobile communications: email, Internet access, secure VPN, telephony, VoIP, SMS, videoconferencing, live video streaming, file transfer, time critical data transfer, remote surveillance.

The Inmarsat IsatPhone represents a cell phone option at low cost. The service is initially available in Asia, Africa and the Middle East.

FleetBroadband is the tradename of the BGAN technology when it is implemented in the maritime field.

SwiftBroadband is the tradename of the BGAN technology when it is implemented in the aeronautical field (Onair and Aeromobile are using Inmarsat for aircraft communications).

Inmarsat Ventures Plc (Inmarsat) did not report having the roaming agreements with other operators except of SMS interconnection.

Inmarsat and KPN have agreed that all SCCP traffic towards Inmarsat is routed via KPN. For this purpose, the INMARSAT global title range for SCCP routing is linked to the KPN network. The KPN and Inmarsat networks are interconnected for the distribution of SCCP traffic.

This document will describe the number of ranges involved, translation rules, platform releases and contact persons. Note that this document only relates to SCCP traffic for SMS, since Inmarsat has not implemented any roaming due to both terminal and air-interface incompatibility with other terrestrial-based GSM/3G operators.

It is not required for a network operator to have an AA.19 interworking agreement with either KPN or Inmarsat for sending SMS traffic towards Inmarsat subscribers.

9.6 Ground Implementation of the Number Continuity

9.6.1 Case of Standard GSM Core Network

Globalstar case (two core networks: IS-41 and GSM).

In [1.1], Chapter 2, one finds the very detailed message flow for a classical single or multi-IMSI virtual roaming service implementation. Figure 9.8 is the signaling flow for Globalstar's GSM-based handsets, which use a GSM core network. It would be the same thing for Iridium, Inmarsat or Thuraya which also have a GSM core network. The Roaming Hub has a table to correspond the "IMSI auxiliary" (received from the satphone) to the "IMSI nominal" (the GSM).

In (1), the VLR sends a SEND AUTHENTICATION to request an authentication "challenge". To secure the system, this is *looped back (2) to the Globalstar HLR,* which returns a Challenge (RAND and SRES) back to the satphone (3) and (4) through the Roaming Hub. If the IMSI auxiliary in the satphone computes and matches the challenge, the signaling will continue with an UPDATE LOCATION req (5) which is transformed by the Roaming Hub (IMSI auxiliary replaced by IMSI nominal). The GSM HLR will transfer in an INSERT SUBSCRIBER DATA the MSISDN (GSM) (7) which is assigned (8) to the satphone.

9.6.2 Case of IS-41 Core Network

It is the same signaling flow as for GSM core networks, except that the Roaming Hub must translate certain IS-41 messages to their GSM counterpart. Figure 9.9 follows the same pattern and gives a quick equivalence (MIN is the IS-41 equivalent of IMSI).

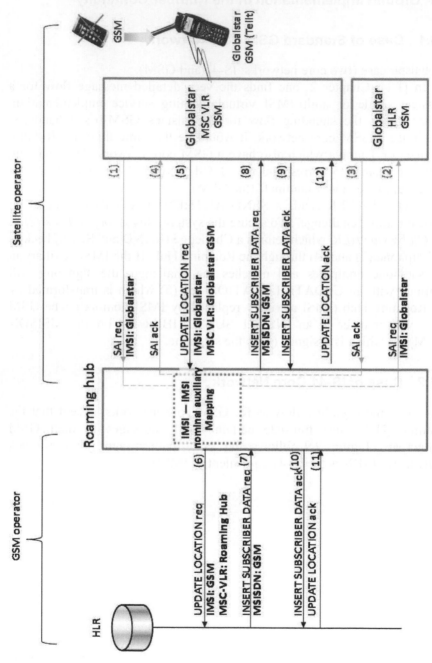

Figure 9.8 Case of a satellite operator GSM core network.

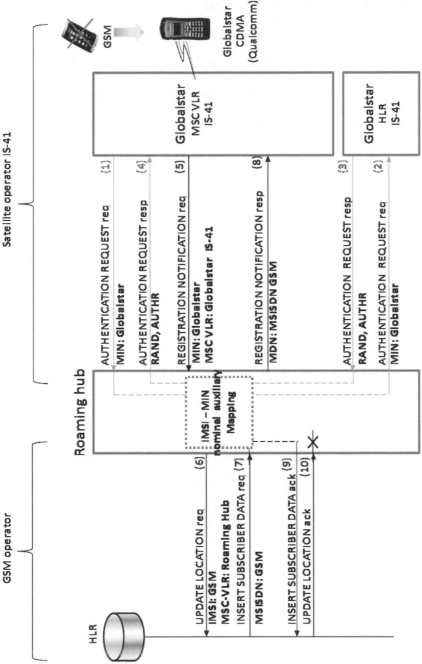

Figure 9. 9 Case of a satellite operator IS-41 core network.

References and Further Reading

[9.1] "Wireless Networks", P.Nicopolitidis, M.S.Obaidat, G.I.Papadimitriou, A.S.Pomportsis, Wiley, 2003, Chap. 7.
[9.2] "Réseaux GSM-DCS", X.Lagrange, P.Godlewski, S.Tabbane, Hermès, 1997, Chap. 5. and 10 ("Handover") .
[9.3] Wolfram Mathematica website, integrals.wolfram.com (beware: gives a real value function only for $|e| > 1$)
[9.4] Maurice Roy, "Mécanique",Vol I Corps rigides, Vol II Milieux continus", Dunod, 1965, Vol I, pp 62-65

10

CDMA (IS-41) <-> GSM Seamless number continuity and application to Globalstar

The quoting mania is our worst enemy

−*Vladimir Ilyitch Ulyianov*

10.1   History and GSM<->CDMA protocols comparison

In 2004, a CDMA->GSM number continuity service was provided by Worldcell (USA), mainly for government officials having a CDMA number phone. When they were going abroad, there was no roaming possible, and they had a GSM phone with many roaming agreements rented for this purpose. The number continuity platform developed by Logica, allowed them to receive calls and SMS on their usual US number, and when they were making calls or SMS their usual US number appeared as CLI. Since then the platform was purchased, but not maintained and is no longer operational.

The number continuity project with Globalstar gave a strong reason to redevelop the technology using a more modern Roaming Hub platform, as Globalstar has two types of core networks and terminals (GSM with an Alcatel HLR in Toulouse, and IS-41 (CDMA) with a DSC HLR in Texas.

It is the same system which would allow, for example, an ordinary CDMA subscriber (example SPRINT in the US) to visit Russia, rent a GSM phone if his own handset is not bi-standard (such as certain iPhones), and get a local IMSI. If this HPLMN has a CDMA<->GSM roaming hub, he would have the full number continuity service.

The CDMA<->GSM Hub is still useful as there are many (> 10% of the world mobile users) CDMA networks in the US (notably SPRINT, Verizon, Metro PCS, Cricket), Asia and Africa (the reason being that the CDMA operators' licenses and the core networks are much cheaper). There exist GSM networks in North America (ANSI) and Europe (ITU),

CDMA networks in North America (ANSI) and Europe (ITU). Table 10.1 gives the differences between ANSI and ITU networks whether they are CDMA or GSM.

10.1.1 TCAP ITU and TCAP ANSI Comparison

However, most GSM networks (T-Mobile USA, AT&T, Canadian GSM) even if they are in an ANSI area use TCAP ITU.

The TCAP ANSI and ITU look similar but are not compatible. It is not just a simple matter of changing the TCAP operation codes, the component codes and the transaction Ids also need to be changed. So if interworking needs to be performed between two networks, one with TCAP ANSI, the other TCAP ITU, the Roaming Hub needs two TCAP instances running in parallel.

Table 10.1 ITU-ANSI differences

	ITU	ANSI
TCAP Layer Transaction codes Component codes Origin and Destination Transaction IDs	TCAP ITU Unidirectional = 61 Hex BEGIN code = 62 Hex CONTINUE = 65 Hex END = 64 Hex Abort = 67 Hex does not exist does not exist Invoke = A1 Hex Return Result = A2 Hex Return Error = A3 Hex Reject = A4 Hex	TCAP ANSI Unidirectional = E1 Hex Query with permission = E2 Hex Continue with permisssion = E5 Hex Response = E4 Hex Abort = 76 Hex Query without permission= E3 Hex Continue without permission=E6 Hex
SCCP Layer	SCCP ITU	SCCP ANSI
Network layer: GT formats	Same	Same
Network layer: Point Codes	14 bits	24 Bits
Nework layer SubSystemNumbers SSN	Same (6=HLR,7=VLR) except SMSC GW = 8	Same (6=HLR,7=VLR) except SMSC = 11

10.1.2 MAP GSM and MAP IS-41 Comparison

Both mobility protocols are called MAP. Table 10.2 gives the most striking differences. We have given the full list regarding the number continuity service, including voice and SMS services.

Table 10.2 GSM<-> IS-41 differences

	GSM	IS-41 (used for CDMA)
Name of Mobily protocol	MAP GSM (3gpp TS 29.002)	MAP IS-41 (TIA/EIA IS-41 D)
Authentication(VLR<->HLR)	SEND AUTHENTICATION INFORMATION req	AUTHENTICATION req
Registration Circuit Services (VLR<->HLR)	UPDATE LOCATION	REGISTRATION NOTIFICATION req
	INSERT SUBSCRIBER DATA	REGISTRATION NOTIFICATION resp
Incoming call to subscriber's number (GMSC-> HLR then HLR->VMSC	SEND ROUTING INFO req	LOCATION REQUEST req
	PROVIDE ROAMING NUMBER req	ROUTING REQUEST req
Deregistration by user (VLR->HLR)	PURGE MS req	MS INACTIVE req
Incoming SMS to subscriber's number (SMSC->HLR then SMSC->VMSC)	SEND ROUTING INFO FOR SM req	SMS REQUEST req
	MT FORWARD SM req	SMS POINT TO POINT DELIVERY req
Outgoing SMS from subscriber (VMSC->SMSC)	MO FORWARD SM req	SMS POINT TO POINT DELIVERY req

	GSM	IS-41 (used for CDMA)
Change of Visited MSC (HLR->old VMSC)	CANCEL LOCATION req	CANCELLATION req
USSD services	PROCESS USSD REQUEST req USSD REQUEST req USSD NOTIFY req	No USSD services in IS-41!!
Data services(Internet)	Circuit mode obsolete, uses ISUP and V110 modems with a IWF	Circuit mode services only in IS-95.See [10.4]
	GTP protocol (see chapter 3)	GTP protocol (CDMA Packet services in CDMA2000). See [10.4] and [10.5]
Registration Packet Services (VLR<->HLR)	UPDATE LOCATION GPRS	REGISTRATION NOTIFICATION req ADD SERVICE[10.4] DROP SERVICE[10.4]
Subscriber public number	MSISDN	MDN for outgoing calls, outgoing SMS and incoming SMS DGTSCAR for incoming calls(GMSC->HLR)
Mobile Subscriber Identity (in the SIM card(GSM) or in the handset (CDMA)	IMSI	MIN
SMS 7 bits alphabet (text coded in 7 bits is not compatible at all between GSM an IS-41)	3gpp TS 23.038, the 7 bits characters are inside an 8 bits format, with every 8 character filled in the first bit of the 8 bits format.	TIA/EAI-637-A, the 7bits characters are simply packed one after the other in a bit string

As a consequence of the Sub System Number (SSN) being the same in GSM and IS-41, and of a common international gateway being used by a service provider of number continuity, a routing of the incoming traffic to the MAP GSM stack or the MAP IS-41 stack cannot be based on the SSN as in most network equipments software (146 goes to Camel, the others to MAP).

For a general operation, there must be two routing levels based on a table of Global Titles (GT) specifying the ANSI networks (GSM or CDMA) (so the incoming traffic is sent to TCAP ANSI or the TCAP ITU), and after the TCAP layer, a table specifying the MAP IS-41 or the MAP GSM. A diagram is given in Figure 10.1.

Such a mixed GSM<->CDMA roaming platform is then much more complex than the implementations which have appeared in the past years, as they necessitate a non-standard SS7 architecture using ITU, ANSI, GSM, IS-41 components with some non-standard routing levels between the layers. The details given below are for those who want to develop or just need to understand how it works.

To simplify, we have assumed that the GSM<->CDMA Hub is connected to an ITU SS7 provider which is offering the ANSI<-> ITU Point Code conversion in the path to the ANSI networks. This is why we see a single MTP3 and M3UA layer as well as a single SCCP ITU layer. If there is no

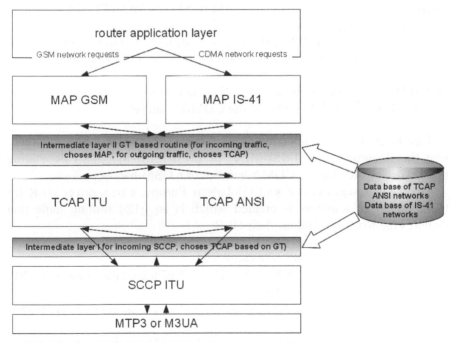

SS7 network with ANSI<->GSM Point Code conversion

Figure 10.1 Architecture of a GMS<->CDMA protocol converter.

ANSI<->ITU Point Code conversion, it is possible to run with two instances of SCCP and two instances of MTP3 or M3UA.

10.2   Rerouting of Registration to the GSM<->CDMA Converting Roaming Hub

Figures 9.8 and 9.9 of the previous chapter show how the registration messages reach the Roaming Hub, that is, SEND AUHENTICATION INFO and UPDATE LOCATION (GSM HLR handsets), AUTHENTICATON REQUEST and REGISTER NOTIFICATION (CDMA HLR handsets).

As an example, the E164 numbering plan for Globalstar GSM Europe is:

33640044200-44999 for CDMA handsets,
33640000000-19999 for GSM handsets,

There are also some ranges for the IMSI (GSM) or MIN (CDMA) assigned to Globalstar Europe.

A sub range of IMSI and MIN is assigned by Globalstar to some planned handsets for the number continuity service. For example:

+208059990040000-49999 for the GSM Handsets
+40379810000-40379819999 for the CDMA Handsets

and as the subscribers subscribe to the service, their IMSI GSM is entered into the Roaming Hub (IMSI-MIN or IMSI-IMSI mapping) depending upon whether they have a CDMA handset or a GSM handset.

In the Gateways GMSCs of Globalstar Europe, a new proxy HLR for these ranges of numbers is created which is an E124 routing table that declares the Roaming Hub as their HLR.

GMSC France+208059990040000-49999
                    ->33XXXXXXXX (GT of the single roaming Hub)
GMSC USA+40379810000-19999
                    ->33XXXXXXXX (GT of the single roaming Hub)

This way all the registration messages are forwarded to the Roaming Hub.

This is shown in Figure 10.2 for the rerouting of number continued CDMA   Handsets.

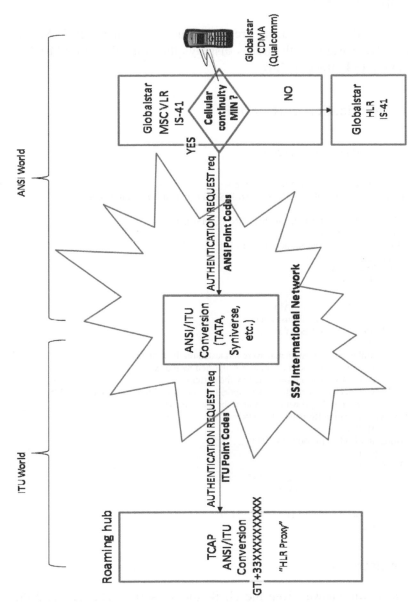

Figure 10.2 Rerouting to the "HLR proxy".

10.3   Details of the GSM<->CDMA Number continuity Implementation

As we assume that most readers are quite familiar with MAP GSM, the protocol analyzer used for the traces displays the equivalence GSM when possible in the IS41 traces below.

10.3.1 Authentication VLR<->HLR

The CDMA handset is powered on.

```
----- IS41 Message Decoding (c)HALYS 2011,2012 –
(28):AUTHENTICATION REQUEST (VLR<->HLR Send Authentication Info in GSM)
Length = 59
          (136):Mobile Identification Number MIN(as IMSI in GSM)
          MIN = +4037990012
          (137):Electronic Serial Number ESN(as IMEI in GSM)
          Manufacturer code = B3
          ESN = B309AFAF
          (149):MSC ID
               7809-10
          (34):System Access Type(SAT)
               (3):Autonomous Registration
          (49):System Capabilities(SYSCAP)
               (0B)
          Authentication parameters were requested on the system access
          Signaling message encryption is supported by the system
          Voice privacy is not supported by the system
          System can execute the CAVE algorithm and share SSD for the indicated MS
          SSD is not shared with the system for the indicated MS
          (35):Authentication Response(AUTHR)
               01387C
          (10):Count Update Record(COUNT)
               (00)
          (32):PC_SSN
               PC(24 bits) = 2247429 SSN= 7
          (40):Random Variable(RAND)
               08016558
          (47):Terminal Type(TERMTYPE)
               (32):IS-95
```

10.3.2 Registration

After a successful answer from the HLR, the subscriber registers and the profile sent by the HLR is loaded in the VLR.

10.3.2.1 VLR->HLR request

----- IS41 Message Decoding (c)HALYS 2011,2012 --
(13):REGISTRATION NOTIFICATION (VLR->HLR Update Location or HLR->VLR
Insert Subscriber Data in GSM)
Length = 59
           4097 (136):Mobile Identification Number MIN(as IMSI in GSM)
           MIN = +4037990012
    (137):Electronic Serial Number ESN(as IMEI in GSM)
        Manufacturer code = B3
        ESN = B309AFAF
    (145):Qualification Information code(QUALCODE)
        (3):Validation and profile
    (150):System My Type Code(MYTYPE)
        (16):QUALCOMM
    (149):MSC ID
        7809-10
    (32):PC_SSN
        PC(24 bits) = 2247429 SSN= 7
    (104):SMS Address(as Visited MSC GT in GSM)
        Type of digit 00
        Nature of number 01
        International
        Presentation allowed
        Number is not available
        (2):Telephony Numbering E164
        (1):BCD
           Number of digits 11
           +16139889998
    (53):Extended MSC Identification Number(EXTMSCID)
        7809-200
    (49):System Capabilities(SYSCAP)
        (0B)

Authentication parameters were requested on the system access
Signaling message encryption is supported by the system
Voice privacy is not supported by the system
System can execute the CAVE algorithm and share SSD for the indicated
MS. SSD is not shared with the system for the indicated MS.

10.3.2.2 HLR->VLR response with the subscriber's profile

----- IS41 Message Decoding (c)HALYS 2011,2012 --
(13):REGISTRATION NOTIFICATION (VLR->HLR Update Location or HLR->VLR
Insert Subscriber Data in GSM)
Length = 65
    (150):System My Type Code(MYTYPE)
        (63):Globalstar
    (142):Authorization Period(AUTHPER)
        (6):Indefinite,value = 0

```
(149):MSC ID
      7808-222
(78):Authentication Capabilities
      (1):No authentication required
(153):Calling Features Indicator(as Call Forwarding Conditions in GSM)(CFI)
      CFNA CFB CFU CD CNIR CNIP1
(93):Mobile Directory Number MDN(as MSISDN for the SRI_FOR_SM or in the
INSERT SUBSRIBER DATA GSM)
      Type of digit 05
      Nature of number 31
      International
      Presentation allowed
      Number is not available
      (2):Telephony Numbering E164
            (1):BCD
            Number of digits 10
            MDN = +4037990012
(151):Origination Indication(ORIGIND)
      (7):International
(152):Termination Restriction Code(TERMRES)
      (2):Unrestricted
(48):CDMA Service Option List(IS-737)(CDMALIST)
      9F812F0200029F812F0202019F812F020101
```

The MDN (the subscriber's number which shows in the calls or SMS is then set by the HLR (exactly as in GSM). The test which was done with Globalstar has a small particularity: the MDN Mobile Directory Number (MSISDN in GSM) is the same as the MIN Mobile Identity Number (IMSI in GSM). In general, it is not the case with other CDMA networks. Also, the CDMA MDN does not include the country code (+1).

In Figure 9.10 of the previous chapter showing the number continuity GSM-> Globalstar CDMA, the system will set the MDN sent to the Globalstar VLR = the GSM MSISDN. So when a call or SMS is made with the Globalstar, the GSM number will show.

10.3.3 Incoming call to CDMA subscriber

In Figure 10.3, a call is made to the MSISDN of the GSM. The GSM HLR will send a PROVIDE ROAMING NUMBER (which includes the GSM IMSI) to the visited VLR which is the Roaming Hub. The Roaming Hub will map the IMSI to the MIN and will send a ROUTING REQUEST including the MIN (same as the IMSI in GSM). This ROUTING REQUEST also has the GT of the GMSC (which is the GT of the Roaming Hub).

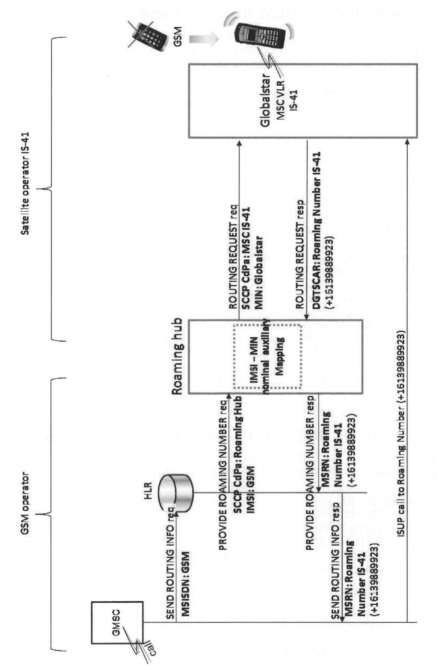

Figure 10.3 Incoming call to CDMA subscriber.

10.3.3.1 ROUTING REQUEST Request (Roaming Hub->VLR)

----- IS41 Message Decoding (c)HALYS 2011,2012 --
(16):ROUTING REQUEST (HLR->VLR Provide Roaming Number GSM)
Length = 52
        (129):Billing ID(BID)
           1E81C918587200
        (136):Mobile Identification Number MIN(as IMSI in GSM)
           MIN = +4037990012
        (137):Electronic Serial Number ESN(as IMEI in GSM)
           Manufacturer code = B3
           ESN = B309AFAF
        (149):MSC ID
           7809-201
        (150):System My Type Code(MYTYPE)
           (16):QUALCOMM
        (47):CDMA Service Options(IS-737)(CDMASO)
           0002
        (94):MSC Identification Number(as GT of GMSC in GSM)(MSCIN)
           Type of digit 00
           Nature of number 31
           International
           Presentation allowed
           Number is not available
           (2):Telephony Numbering E164
               (1):BCD
               Number of digits 11
               +33XXXXXXXXX /* GT of Roaming Hub */

10.3.3.2 ROUTING REQUEST Response (VLR->Roaming Hub)

----- IS41 Message Decoding (c)HALYS 2011,2012 --
(16):ROUTING REQUEST (HLR->VLR Provide Roaming Number GSM)
Length = 37
        (149):MSC ID
           7809-10
        (129):Billing ID(BID)
           1E810A65CF3C00
        (132):Digits(called MSISDN from GMSC or MSRN returned by VLR in GSM)(DGTSCAR)
           Type of digit 06
           Nature of number 01
           International
           Presentation allowed
           Number is not available
           (2):Telephony Numbering E164
               (1):BCD
               Number of digits 11

+16139889923 /* Roaming Number allocated by the VLR IS-41 and
returned to the Roaming Hub */
(32):PC_SSN
        PC(24 bits) = 2247429 SSN= 7

The Roaming Hub will give +16139889923 in the GSM PROVIDE
ROAMING NUMBER Confirmation. As a result, the GSM GMSC will
call +16139889923 directly and the IS-41 network will receive the same
Mobile Terminated Calls charges.

10.3.3.3 Call Forwarding IS-41 for Unsuccessful Mobile Terminated Calls

This does not work like GSM. In GSM, the VLR profile contains
"conditional call forwarding" information for call busy, no response, not
reachable. There is no such thing in IS-41, the profile returned by the HLR
in the REGISTER NOTIFICATION result returned does not have it. When
the incoming call of Figure 10.3 fails, the VLR IS-41 sends to the HLR a
TRANSFER TO NUMBER REQUEST with the "Redirection Reason",
asking for instructions.

----- IS41 Message Decoding (c)HALYS 2011,2012 --
(23):TRANSFER TO NUMBER REQUEST (VLR->IILR->VLR)(the VLR tells the result
of a MT call and receives a redirection number)
Length = 22
  (136):Mobile Identification Number MIN(as IMSI in GSM)
    MIN = +4037990012
  (137):Electronic Serial Number ESN(as IMEI in GSM)
    Manufacturer code = B3
    ESN = B309AFAF
  (150):System My Type Code(MYTYPE)
    (16):QUALCOMM
  (147):Redirection Reason
    (4):No Page Response

The HLR responds by sending a *"Redirecting Number"* which could be the
GSM VMS number, which is then called by the VLR. For the number
continuity service, the Roaming Hub has extracted the GSM conditional
call forwarding numbers from the INSERT SIBSCRIBER DATA and uses
them to create the TRANSFER TO NUMBER Response sent to the VLR.

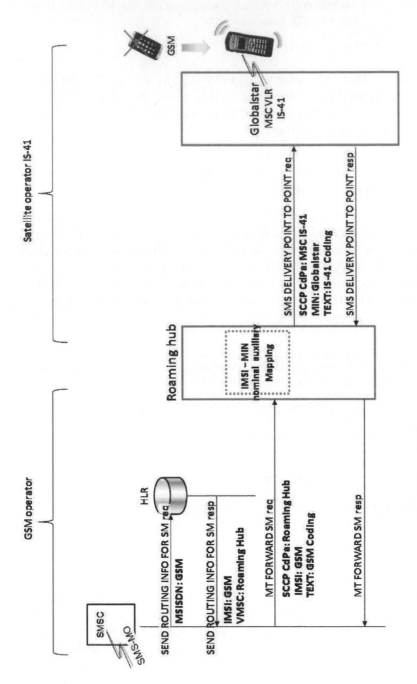

Figure 10.4 Incoming SMS-MT to CDMA subscriber.

----- IS41 Message Decoding (c)HALYS 2011,2012 --
(23):TRANSFER TO NUMBER REQUEST (VLR->HLR->VLR)(the VLR tells the result of a MT call and receives a redirection number)
Length = 36
 (132):Digits(called MSISDN from GMSC or MSRN returned by VLR in GSM)(DGTSCAR)
    Type of digit 01
   Nature of number 00
    National
    Presentation allowed
    Number is not available
  (2):Telephony Numbering E164
  (1):BCD
  Number of digits 10
 +4037990012
(96):No Answer Time
 0 seconds
(100):*Redirecting Number* Digits(number where call is forwarded(e.g.VMS)
   Type of digit 01
   Nature of number 01
    International
    Presentation allowed
    Number is not available
  (2):Telephony Numbering E164
  (1):BCD
  Number of digits 11
 +33609000123   /* GSM VMS number */
 (122):Termination Triggers
  (0):BUSY:Busy
  (4):RF:Failed call
  (8):NPR:No Page Response Call
  (12):NR: Member not reachable

The VLR will then forward the voice call to +33609000123 as shown in Figure 10.3.

10.3.4 Deregistration

This occurs when a subscriber powers down the handset. A signaling message is sent from the VLR to the HLR to deregister the handset.

----- IS41 Message Decoding (c)HALYS 2011,2012 --
(22):MS INACTIVE (VLR->HLR MS Purge GSM)
Length = 16
       (136):Mobile Identification Number MIN(as IMSI in GSM)
         MIN = +4037990012
       (137):Electronic Serial Number ESN(as IMEI in GSM)
         Manufacturer code = B3
         ESN = B309AFAF

10.3.5 Incoming SMS-MT to CDMA Subscriber

10.3.5.1 The Local or Foreign SMSC Asks the HLR for the Visited MSC and MIN

This is the case of a SMSC, not the "number continuity" case which is simpler and does not need to interrogate the HLR IS-41.

The local or foreign SMSC uses the known public number MDN of the subscriber same as the MSIDN in a SRI_FOR_SM and gets the Visited MSC GT and the MIN (same as IMSI) which will be used in the SMS DELIVERY POINT TO POINT (same as FWD_SM_MT in GSM).

```
----- IS41 Message Decoding (c)HALYS 2011,2012 --
(55):SMS REQUEST (SMSC<->HLR Send Routing Information for SM in GSM)
Length = 24
        (109):SMS Notification indicator(demand of a HLR alert by the
        SMSC)(SMSNOTIND)
            Notify when available(01)
        (116):SMS Teleservice Identifier(IS-637)(SMSTID)
            CDMA Number Messaging Teleservice(1002)
        (93):Mobile Directory Number MDN(as MSISDN for the SRI_FOR_SM or in the
        INSERT SUBSRIBER DATA GSM)
            Type of digit 05
            Nature of number 31
            International
            Presentation allowed
            Number is not available
            (2):Telephony Numbering E164
                (1):BCD
                Number of digits 10
                MDN = +4037990012
```

This is the response of the HLR including the MIN. This is particular (Globalstar) and they use a MDN (the MSISDN in GSM) equal to the MIN (the IMSI in GSM).

```
----- IS41 Message Decoding (c)HALYS 2011,2012 --
(55):SMS REQUEST (SMSC<->HLR Send Routing Information for SM in GSM)
Length = 29
        (137):Electronic Serial Number ESN(as IMEI in GSM)
            Manufacturer code = B3
            ESN = B309AFAF
        (104):SMS Address(as Visited MSC GT in GSM)
            Type of digit 05
            Nature of number 31
            International
            Presentation allowed
            Number is not available
            (1):ISDN Numbering Plan
                (2):IA5 International Alphabet 5
                Number of digits 11
```

                    +16139889998
        (136):Mobile Identification Number MIN(as IMSI in GSM)
            MIN = +4037990012

10.3.5.2 The SMSC Sends the SMS to the Visited MSC

For number continuity, the Roaming Hub did not need to interrogate the HLR IS-41 because it already knows the MIN and the visited VLR IS-41. It will send the SMS DELIVERY POINT TO POINT directly.

----- IS41 Message Decoding (c)HALYS 2011,2012 --
(53):SMS DELIVERY POINT TO POINT (SMSC->MSC SMS-MO or MSC->SMSC SMS-MT Forward Short Message GSM)
Length = 188
        (136):Mobile Identification Number MIN(as IMSI in GSM)
            MIN = +4037990012
        (137):Electronic Serial Number ESN(as IMEI in GSM)
            Manufacturer code = B3
            ESN = B309AFAF
        (105):SMS Bearer Data
                MESSAGE_ID:
                message type= 01(Deliver(mobile terminated only))
                message_ID: 0AD7
                USER_DATA
                Subparam length=16
                Msg_Encoding= 02(7 Bits ASCII)
                Num 7b characters =16
                User_Data= Globalstar
                NUMBER OF MSGs IN VMS: 12
                LANGUAGE_INDICATOR: 02(French)
                MESSAGE CENTER TIMESTAMP: year= 2011 month= 12 day= 10
                hour= 6 min= 45 sec= 32
                VALIDITY PERIOD(absolute format): year= 2011 month= 12 day=
                14 hour= 10 min= 0 sec= 0
                VALIDITY PERIOD(relative format)=85(that is 25800 seconds)
                ALERT ON MESSAGE DELIVERY: Use high priority alert
                MESSAGE DISPLAY MODE(1): Mobile default setting: as
                predefined in the MS
                REPLY OPTION:
                User Ack(if this is SMS-MT)=Positive(manual) User ACK requested
                from the recipient user
                Delivery Ack(if this is SMS-M0)=No Delivery ACK requested from
                the recipient
                PRIORITY INDICATOR(2): Urgent
                PRIVACY INDICATOR(3): Secret
                DEFERRED DELIVERY TIME(absolute format): year= 2012
                month= 1 day= 6 hour= 23 min= 59 sec= 0
                DEFERRED DELIVERY TIME(relative format)=84(that is 25500
                seconds)
                USER RESPONSE CODE(predefined by SMSC for the SMSack)=33

CALL BACK NUMBER
Digit Mode= 00(DTMF (4 bits BCD))
Num_Fields=13
Call Back Number= 1234567890ABC
CALL BACK NUMBER
Digit Mode= 01(ASCII (8 bits))
Numbering type=01
Numbering plan=02
Num_Fields=6
Call Back Number= 1234AB
(116):SMS Teleservice Identifier(IS-637)(SMSTID)
CDMA Voice mail notification(4099)
(109):SMS Notification indicator(demand of a HLR alert by the
SMSC)(SMSNOTIND)
Notify when available(01)

10.3.5.3 The HLR Alerts the SMSC When Subscriber Becomes Reachable

This is like ALERT SERVICE CENTER in GSM. The SMSC will retry.

---- IS41 Message Decoding (c)HALYS 2011,2012 --
(54):SMS NOTIFICATION (HLR->SMSC Alert SC GSM)
Length = 29
(137):Electronic Serial Number ESN(as IMEI in GSM)
Manufacturer code = B3
ESN = B309AFAF
(136):Mobile Identification Number MIN(as IMSI in GSM)
MIN = +4037990012
(104):SMS Address(as Visited MSC GT in GSM)
Type of digit 05
Nature of number 31
International
Presentation allowed
Number is not available
(1):ISDN Numbering Plan
(2):IA5 International Alphabet 5
Number of digits 11
+16139889998

10.3.6 Internet Data Services for CDMA->GSM Number Continuity

The likely usage is a GSM usage in a VPLMN. The Local Break-Out of Chapter 3 provides a simple solution so that the CDMA subscriber can have the internet access while using a GSM handset. But the GTP Protocol is common for GSM and CDMA2000, so that the PDP Context can also be established with the HPLMN CDMA GGSN [10.5].

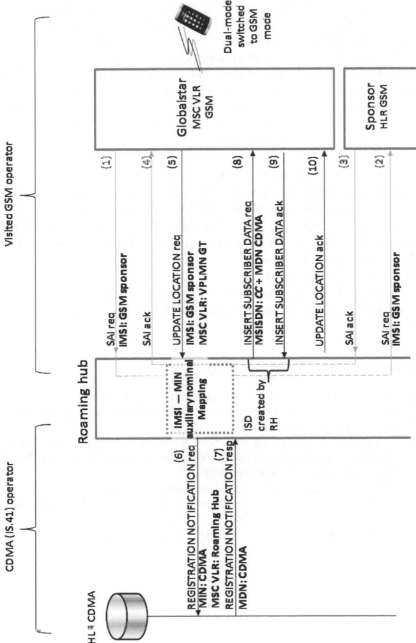

Figure 10.5 CDMA<->GSM registration with a Roaming Hub.

10.4 CDMA<->GSM Number Continuity Service

This is a real case with major networks, such as SPRINT(CDMA) in the USA which provide the international roaming services to their subscribers with a dual standard handset CDMA + a GSM SIM card from a "sponsor" and a Roaming Hub supplier providing the conversion. This is not a new idea and the service was offered since 2003 with a separate GSM handset provided by a sponsor. It is more practical with dual standard handsets including the latest iPhone. When he arrives in Europe for example, the user selects the GSM mode or uses a separate GSM handset, the number continuity service is provided with the IS-41 HLR thanks to the two-way conversion in the Roaming Hub as shown in Figure 10.5. Compare with the Figures 9.8 and 9.9. In IS-41 there is no equivalent of MAP GSM INSERT SUBSCRIBER DATA, the MDN of the user (equivalent of the MSISDN) is sent by the HLR CDMA in the REGISTRATION NOTIFICATION resp (7). The Roaming Hub creates an INSERT SUBSRIBER DATA req (8) which contains the MSISDN to be used in GSM roaming. Figure 10.5 shows
MSISDN = CC + MDN CDMA
CC would be +1 in the case of a US roamer, which is added by the Roaming Hub because in many cases the MDN does not include the country code CC.

References and Further Reading

[10.1] TIA/EIA-41-D, "Number Radiotelecommunications Intersystem Operations", 1997 (main description of the IS-41 protocol with the tables of operation and parameter codes)
[10.2] TIA/EIA-637-A, "Short Message Service for Spread Spectrum Systems", 1999 (complement to [10.1] for the SMS service. Gives all the details and tables to implement the SMS service)
[10.3] TIA/EIA-737, (describes additional parameters to [10.1])
[10.4] TIA/EIA-707-D, "Based Network Enhancements for CDMA Packet Data Service(C-PDS)",3GPP2 N.S0029 V1.0.0, june 2002.
[10.5] IFAST#24/2004.10.04/07, "CDMA "Packet Data Roaming eXchange guidelines"

11

Anti-Steering of Roaming System with Roaming Hubs

The one century armor-big gun battle started in 1860 with the armored frigate "Gloire" designed by famous naval engineer Henri Dupuy de Lôme and finished in 1952 with the completion of the last battleship, the great "Jean Bart".
*   −Henri Maine*

11.1 Principle of The Passive "Steering of Roaming"

For a mobile operator, it is interesting to choose networks visited by his outbound roaming subscribers, in order to have the best tariffs, in particular using their foreign subsidiaries. The GSMa considers this a right of the HPLMNs which no one will object, provided that the technical way to achieve this is coherent with respect to the "SS7 stack" standard MAP, TCAP, SCCP and lower layers, and does not create undue charged SS7 traffic in the "steered out" foreign partner VPLMNs, which are often disarmed against this "steering".

The right way to do this "steering" is the dynamic setup of preferences in the SIM cards of the outbound roamers [11.3]. In this case, the roamer will always attempt first a preferred network in the EFplmnsel file of his SIM card. The desired statistical distribution of the usage of the various partner VPLMNs of a country is achieved by following that distribution with preference setups not identical for all SIM cards.

The much used "passive" SS7 steering is depicted in Figure 11.1 with the HPLMN and with steering on the left and the VPLMN on the right.

The steering system is "passive": it monitors the incoming UPDATE LOCATION Request (2) but is able to inject extraneous "confirmations" such as (5).

The mobile attempts to register in the VPLMN which sends a normal standard UPDATE LOCATION Req (1) then (2) to the partner HLR, which accepts and starts sending the profile in INSERT SUBSCRIBER

203

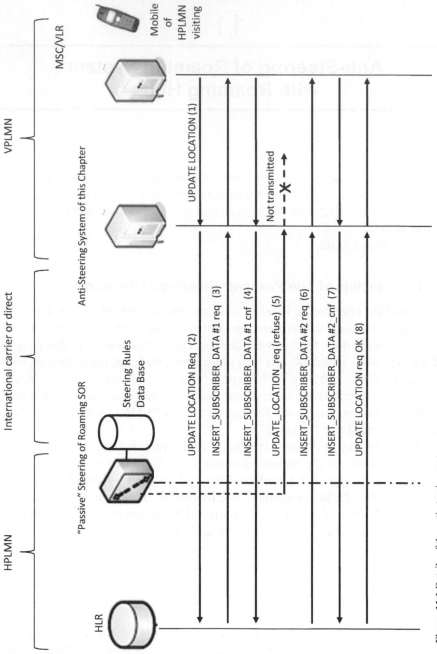

Figure 11.1 Details of the anti-steering system.

DATA (3), (6) which would be Acked normally (4), (7) without the action of the steering. Then the HLR would send the UPDATE LOCATION Cnf (8) and the visitor could use the VPLMN.

The steering of roaming as soon as it sees the UL (1) would decide to let the registration proceed or forbid it in which case it sends a UPDATE LOCATION Cnf (5) with an error code (35 and 36 are used mostly by SoR vendors), in a TCAP END message. The receiving MSC-VLR interrupts then the overall TCAP transaction that it has initiated. The following INSERT SUBSRIBER DATA and UPDATE LOCATION Cnf (OK) received (6) and (8) are discarded by the MSC-VLR TCAP because the transaction was previously closed. This handset's attempts to register fails and it tries the next visible network. The selection process by the "steering" continues, but of course it should accept when the last of his roaming partners' network in the visited country is tried.

This method creates undue traffic in the rejected VPLMNs and error logs in their TCAP stack, while error 35 ("Data Missing") and 36 ("Incorrect data values") are incoherent because the UPDATE LOCATION was perfectly formatted. But it works for the HPLMN if they do not complain.

11.2 Anti-steering Countermeasure in the VPMLN

11.2.1 Active Architecture to Filter Incoherent "Spoof" SS7 Messages

The VPLMN is equipped with a front-end system managing the SS7 signaling of the international links. The system illustrated by Figure 11.2, can also provide SMS control (anti-spoof, anti-spam) as described in [11.4].

The "anti-steering" acts as TCAP cleaner, provided by a special protocol layer between the Network layer MTP3 (TDM with E1s) or M3UA (Sigtran) and the SCCP GT network layer. It could also be between the SCCP and the TCAP layer, but it must be before the MAP layer because the transactions do not proceed upward in the SS7 stack if they are closed by the reception of a TCAP BEGIN.

The anti-steering system receives the messages from/to the VPLMN and relays them to/from the partner HPLMNs. It is an "active" configuration which also is able to filter very efficiently the spam SMS or any incorrect SS7 signaling message received. It is exactly what it does for the "anti-steering". If the "filter layer" receives a TCAP END from the international network for an UPDATE LOCATION with an error code 35

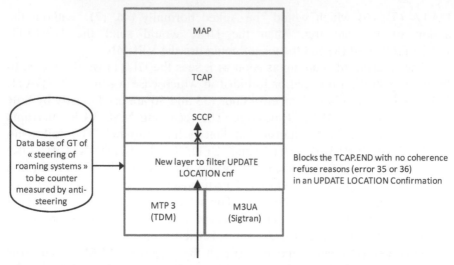

Figure 11.2 Software architecture of anti-steering system (down layer).

or 36, it knows it is incoherent because his network has complaints with standards and the UPDATE LOCATION that it sent was correct. Considering this as "spoof" the "filter layer" discards it. The "filter layer" must keep a context using the TCAP Transaction ID to recognize a "Return Result Last" of an UPDATE LOCATION in a TCAP END because *no Operation Code is included in general in a TCAP END.*

As a result of this filtering, the INSERT SUBSCRIBER DATA (6) will be transmitted to the MAP instead of being dropped, will be Acked (7) and the UPDATE LOCATION Cnf (8) also. The visitor is registered.

11.2.2 Detail of the Anti-Steering Sequence

The trace of Figure 11.3 on the SS7 international links shows exactly how it works.

In line 1 of the trace, the VPLMN has sent a correctly formatted UPDATE LOCATION in a TCAP BEGIN, component type "invoke". It received (line 2) a TCAP END, component type "return error". Using the TCAP Transaction ID, the filter layer finds that the Operation Code is an "UPDATE LOCATION" and the error code 35 is a spoofing attempt because the message sent was correct. This message is dropped. As the partner's HLR was proceeding (it is independent of the "steering system"), the two INSERT SUBSCRIBER DATA are normally received (lines 3 and 5) and Acked (line 6) by the VLR (the transaction TCAP is not closed), so

N°.	SCCP CgPa	SCCP CdPa	TCAP tid	Protocol	Info
1	22C⸺⸻	33⸺7502⸺	00002354	GSM MAP	invoke updateLocation
2	33⸺002021	⸻	00002354	GSM MAP	returnError
3	33⸺002180	⸻	00002354	GSM MAP	invoke insertSubscriberData
4	22C⸻	33⸺002180	C05D60A3	GSM MAP	returnResultLast
5	33⸺002180	⸻	00002354	GSM MAP	invoke insertSubscriberData
6	22C⸻	33⸺002180	C05D60A3	GSM MAP	returnResultLast
7	3368⸺9002180	⸻	00002354	GSM MAP	returnResultLast updateLocation

```
        ▽ Destination Transaction ID
            Transaction Id: 00002354
          oid: 0.0.17.773.1.1.1 (id-as-dialogue)
        ▽ dialogueResponse
            Padding: 7
          ▷ protocol-version: 80 (version1)
            application-context-name: 0.4.0.0.1.0.1.2 (networkLocUpContext-v2)
            result: accepted (0)
          ▷ result-source-diagnostic: dialogue-service-user (1)
          ▷ components: 1 item
    ▽ GSM Mobile Application
      ▽ Component: returnError (3)
        ▽ returnError
            invokeID: 1
          ▽ errorCode: localValue (0)
            localValue: dataMissing (35)
```

Figure 11.3 Trace of an incoherent UPDATE_LOCATION_CNF generated by a SoR system.

that finally the HLR sends a successful TCAP END with a component "return result last".

So, it is quite easy to protect networks (an anti-steering is a bay-product) from the spoof type "passive" steering systems which are the most frequent.

11.3 Anti-Steering Countermeasure Service in a Roaming Hub

A Roaming Hub supplier can easily provide the "anti-steering" (as well as may be the "steering") using preferably the OTA SIM method, for his clients as all the UPDATE LOCATION go through the Roaming Hub much as in Figure 11.1. They are "front end" to their clients and have the possibility of selecting the clients for which the anti-steering service is offered. The marketing dilemma will eventually be handled the same way as do some weaponry merchants: sell both sides.

References and Further Reading

[11.1] Henri Maine, "Histoire de la Marine", 1972, Vol I, II and III

[11.2] B. Mathian, A. Henry-Labordère, "Système de contre-mesure au pilotage de l'itinérance", Patent FR 12 61 791

[11.3] A. Henry-Labordère, "Virtual Roaming Systems for GSM, GPRS and UMTS", Wiley, 2009, Chapter 8, section 8.3.3.6

[11.4] A. Henry-Labordère, V. Jonack, "SMS and MMS interworking in Mobile Networks", Artech House, 2004, Chapter 5, section 5.3

12

Mathematical Models for the Steering of Roaming

How can you candidate as associate engineer in this important project if you cannot in less than 5 minutes prove with elementary plane geometry that the product of two point rotations around a point is a point rotation and build the exact center and the angle with a compass and a rule? A modern engineer must know some mathematics. It would be trivial with the new theory of groups but this matter is not yet taught in our engineering schools. In not so long you will see Laplace's analytical probabilities also included.
      −Gustave Eiffel (1832-1923)

12.1 The Mathematical Model Behind the GLR ("Gateway Location Register"): Why It Helps to Grow the Usage by the Inbound Visitors of a GLR Equipped VPLMN

12.1.1 "Anti Steering" versus "Steering of Roaming" vs "Anti-Anti Steering"

The GLR concept and implementation was introduced a few years ago and is now approved by the GSMa as an authorized "anti-steering of roaming" method.

When a visitor tries to register the first time, it is useful to use an anti-steering system such as the one described in Chapter 11 else he may never register.

The principle of the GLR is that when a visitor registers in a foreign network; his VLR information (the profile) is recorded in the GLR, which monitors the international signaling traffic. Every time he initiates a new UPDATE LOCATION request (UL), instead of forwarding the UL to the HLR of the visitor, the GLR responds, behaving as the HLR, with the VLR information (the GLR sends the "Insert Subscriber data" and the UPDATE LOCATION confirmation).

209

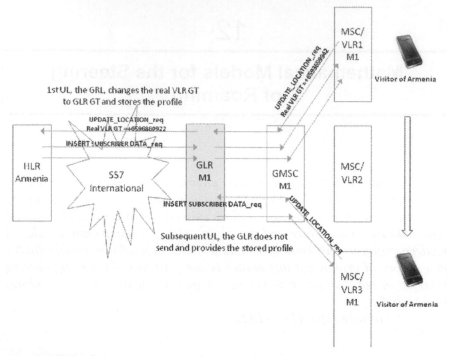

Figure 12.1 GLR Message flow Singapore visited, Armenia visiting.

The official purpose is to save on international SS7 signaling; the real main one is to avoid a new "steering" by the HPLMN, which could give back the customer to a more preferred competitor. In some special cases, such as the Roaming Hub+GLR of an in-flight GSM operator, it is used to block the UL until the travelers of certain airlines companies specifically ask (a USSD transaction) to open the service. This is notably to avoid travelers from being charged unwillingly for incoming calls.

12.1.2 Understanding the Visitor's Handset Selection Process of a Visited Network

All the handsets have the same standard logic, which participates in the VPLMN selection.

When turned on (e.g. landing of an aircraft), they scan the radio channels to scan for MCC-MNC of the active mobile operators.

If the last MCC-MNC, the network which was used (recorded in the SIM card, the EFlocinfo SIM file) is received with the minimum power, the handset will try this one first.

This is in fact, the key to the first "GLR theorem":

> If a VPLMN has a full coverage, with a GLR, it will never lose the visitor once he has used this VPLMN first time.

Also, as anyone who has "tested SIM cards" knows, there is no new UL sent by the VLR if you turn off/on the tested SIM card, one needs to erase manually the profile in the VLR to have a new one. This is the 2nd "GLR theorem".

> If a VPLMN has a single MSC/VLR, there is no utility in a GLR, the unique VLR provides the basic function.

"Basic function" means the VLR will never lose the visitor. However, a GLR with advanced algorithms to help acquire the visitor on his first attempts will speeden up having the visitor in the VPLMN.

If the last MCC-MNC is not available (case of the visitor-in at the airport), if the signal is enough, the handset will:

a) Skip the "forbidden networks" in the SIM card,

b) Select the top "preferred PLMN" in the SIM card, if there is one (this is the best "steering method") and try to establish a signaling channel with the radio network of that operator. The VLR will initiate the UPDATE LOCATION request. If the VPLMN prefers "steering" another network, it will refuse and the handset will try the next network, until if finds one which is also accepted by the HPLMN.

There are two instances of this selection process being started:

1) If the visitor turns off/on (night/morning) his handset.

2) Also, if he was with a competitor which had a coverage hole in an area (signal below threshold), and a better signal above threshold was found, a new selection process is started, with a new chance for this VPLMN, to succeed in registering the visitor with this probabilistic explanation of the efficiency of a standard GLR.

We assume a visitor with N "selection processes" (as above, two in the example)/ day which remains M-days in the country.

Let p (0.3 for the numerical example) be the percentage of visit that the "steering of roaming" HPLMN has allocated to his GLR equipped partner.

12.1.3 What is the Improvement in Percentage of Visitor's Presence in a GLR Equipped VPLMN?

Without a GLR and with the competitors not having a GLR:
   The visitor will use the VPLMN during $p = 30\%$ of his stay.
With a GLR and with the competitors not having a GLR (simpler modeling case):
   Let P_1={Probability that on any given day the visitor registers for the first time in the GLR-equipped VPLMN}
$$P_1 = p + p(1 - p) + p(1 - p)^2 + ... + p(1 - p)^{N-1}$$
which is numerically:
   $P_1 = 1 - (1-p)^N$ which is 0.51 with N=2 (better that 0.3 which is the case without GLR)
Let Pi={Probability that on day # i the visitor newly registers in the GLR equipped VPLMN}.
   We have:

$$P_2 = P_1(1 - P_1) \qquad \text{day 2, 2nd day in the country}$$
$$Pi = P_1(1 - P_1)^{i-1} \qquad \text{day } i$$
....
$$P_M = P_1(1 - P1)^{M-1} \qquad \text{day } M \text{, last day in the country}$$

Probability P that the visitor has registered in the VPLMN during his stay.
   $P = P_1 + P_2 + P_3 + ... + P_M$.
This is a finite geometrical series which sum is the classical result
   $P = 1 - (1 - P1)^M$, giving $P = 0.97$ with M=5 days in the country
Average day number M when the visitor registers successfully first time
   This is the mean M of the random variable i, that is:
   $M = 1P_1 + 2P_2 + 3P_3 + ... + MP_M = P_1 (1 + 2(1 - P_1) + 3(1 - P_1)^2 + M(1 - P_1)^{M-1})$
To compute the value, we use the continuous series results. The sum of the first terms of the continuous series of term $ii(x) = x^i$ is:
   $1 + x + x^2 + x^i + ... + x^M = (1 - x^{M+1})/(1 - x)$
The derivative series $u'i(x)$ has a sum of the first terms which is the derivative of $(1 - x^{M+1})/(1 - x)$:
   $1 + 2x + 3x^2 + ix^{i-1} + Mx^{M-1} = (- (M+1)x^M (1 - x) + (1 - x^{M+1}))/(1 - x)^2$
Using $x=(1 - P_1)$, replacing in M, and simplifying, we have:
   $M = (- (M+1)(1 - P_1)^M P_1 + (1 - (1 - P_1)^{M+1}))/P_1$
Numerically, we have $M = 1{,}76$ days where the visitor becomes "captured by the VPLMN" for a 5 days stay. That is, he uses the VPLMN $U = (5 -$

1.76) / 5 = 65% of his stay instead of only 30% without a GLR function. For longer stays, the function

$U = (M - M)/M$ tends to 1 of course as the duration of the stay grows. Remember the hypothesis:

- The VPLMN has full coverage, so that any time the visitor turns on his handset, he retries on the same VPLMN (yours) because of the SIM card recording of last location. This is the key to the GLR usefulness and is sometimes not explained by the vendors.
- The competitors do not have a GLR.
- We assume a "classical GLR" without the case of anti-steering algorithms (the result could be better of course).

If one of these hypothesis is changed, you may easily build from above the appropriate mathematical model, this is an interesting exercise (case of two competitors in a country, one having also a GLR, should they bury the war axe and both discard their GLR to the displeasure of the vendors?).

Mathematical modeling saves a lot of wasted testing time when a model gives an accurate solution.

For a visitor who has "steering of roaming" by the SIM card, a GLR has few chances of being of any use (only if the preferred competitor has coverage holes which leave a chance to the GLR to acquire the visitor).

12.1.4 Anti-Spoofing SMS-MO Necessary Adaptation in Case of a GLR

As shown by Illustration Figure 12.1, the visitor's HLR (Armenia) has the GLR GT as VMSC GT. If the visitor sends a SMS-MO, the Calling Party SCCP is the real MSC GT. They do not match and the SMSC in Armenia will reject it. To leave the SMS-MO function, the anti-spoofing algorithm must use the 'mask feature' and consider that any GT starting with +65936, +65968, +65976 match (all the "network nodes GT" in the IR21 of M1 Singapore). There is no fixed correspondence between the real VLR GT (several) and the GLR GT (can be unique).

12.2 Mathematical Model for the "Steering of Roaming by SS7"

There are two methods, the most efficient being the automatic setup by OTA SIM of the "preferred PLMN" and also the widely used "Steering by SS7"".

12.2.1 Principle of the "Steering by SS7"

Assume there are N roaming partners available and no "preferred PLMN" in the SIM card. The outbound subscriber's *handset will scan the N visible networks* in an order B, D, A, C which depends on the signal strength and sends each time an UPDATE LOCATION message to his home network (HPLMN) if the previous one is rejected by the Steering of Roaming (SoR) in the HPLMN. The SoR system receives them and by selective rejects of the UL may achieve a desired distribution of usage of the various roaming partners. We give below a description of the optimization problem involved with numerical examples. We consider different algorithms for the SoR.

12.2.2 No Selection Process: Accept the 1rst UL (Number of Draws p=0)

There is a first UL which concerns network A, B, C ... F randomly and it is accepted. We have:

$1/N = P_0(A) = P_0(B) = ... = P_0(F)$ for the probability of having A, B as the VPLMN assuming that the coverages are the same. It means that there is no SoR and the first UL B is accepted.

12.2.3 Simple Selection Process: up to 2 UL Maximum (Number of Draws p=1)

The selection algorithm, whenever an outbound subscriber makes a first UL on network B, makes a random draw of B with probability ($0 \leqq b \leqq 1$). If B is not drawn, there is a second UL which will come randomly from A, C,..F. With the "$p=1$ draw algorithm", we always select this second network.

Share of each roaming partner's network utilization by the outbound subscriber.

We will consider to simplify, an example with $N=3$. Let $P_1(A), P_1(B), P_1(C)$ the value that we want to compute for the probability of use of the networks A, B and C. We assume that all networks have full coverage.

$P_1(A) = 1/N \times a + 1/N \times (1 - b)/(N - 1) + 1/N \times (1 - c)/(N - 1)$.

The third term corresponds to the case of C being the first UL, the system does not draw C and the second UL (accepted) comes from A (with $N=3$, there is $1/(N - 1) = \frac{1}{2}$ probability that this is the case.
 With simplification, we write:

$$a + 1 - (b+c)/(N - 1) = N P_1(A)$$
$$b + 1 - (c+a)/(N - 1) = N P_1(B) \qquad (12.1)$$
$$c + 1 - (a+b)/(N - 1) = N P_1(C)$$

which allows to compute the probabilities $P_1(A)$, $P_1(B)$, $P_1(C)$ knowing a, b, c and $N=3$.
 One verifies that:
$P_1(A)+P_1(B)+P_1(C) = 1/N \times (a + b + c + N - (N - 1)(a + b + c)/(N - 1)) = 1$

and also that these N equations are not independent, we see by adding the N equations (1) that the sum of the coefficients of a, b and c are all 0, so we have:
 $N = N(P_1(A)+(P_1(B)+P_1(C))$ which is always true whatever a, b, c

12.2.4 Selection Process With More Attempts: up to 3 UL Maximum (p=2)

A more complicated example with $N=4$ and draw probabilities a, b, c and d.
 If the first draw does not select the incoming UL, we make up to two attempts and always accept the third. As expected, by setting proper values of a, b, c and we can guarantee get $P_2(A) = 1$ for a given Roaming Partner's network if there are three such networks. The following possible sequences selecting A are:
A   1st UL from A (selected)

N-1 cases where A is selected after 2 UL:
B A   1st UL from B (not selected), 2nd UL from A (selected)
C A   1st UL from C (not selected), 2nd UL from A (selected)
D A   1st UL from D (not selected), 2nd UL from A (selected)

(N-1)(N-2) cases where A is selected after 3 UL:

B C A   1st UL from B (not selected), 2nd UL from C (not selected), last UL (p=2) from A (no draw)

B D A   1st UL from B (not selected), 2nd UL from D (not selected), last UL (p=2) from A (no draw)

C B A   1st UL from C (not selected), 2nd UL from B (not selected), last UL (p=2) from A (no draw)

C D A   1st UL from C (not selected), 2nd UL from D (not selected), last UL (p=2) from A (no draw)

D B A   1st UL from D (not selected), 2nd UL from B (not selected), last UL (p=2) from A (no draw)

D C A   1st UL from D (not selected), 2nd UL from C (not selected), last UL (p=2) from A (no draw)   (12.2)

It gives:

$P_2(A) = 1/Nxa$ (select A on 1st UL)
$+ 1/N(N - 1)x((1 - b)+(1 - c)+(1 - d))xa$ (select A on 2nd UL)
$+ 1/N(N -1)(N -2)x((1-b)(1-c)+(1-b)(1-d)+(1-c)(1-b)+(1-c)(1-d)+(1-d)(1-b)+(1-d)(1-c))$ (select A on 3rd UL)
                                                          (12.3)

Multiplying both sides of (12.3) by $N(N - 1)(N - 2)$, while making the formula general for more than $N=3$, then simplifying, it gives:

N(N-1)(N-2)P_2(A)= (N-1)(N-2)**a** + ((N-1)(N-2)**a**-(N-2)(**ba+ca+da**) + (N-1)(N-2)–2(N-2)(**b+c+d**) +2(**bc+bd+cd**), which gives 4 equations below:

2(N-1)(N-2)**a** -(N-2)(**ba+ca+da**) –2(N-2)(**b+c+d**) +2(**bc+bd+cd**) = N(N-1)(N-2)P_2(**A**) – (N-1)(N-2)

2(N–1)(N–2)**b**– (N–2)(**cb+db+ab**)–2(N–2)(**c+d+a**) +2(**cd+ca+da**) = N(N–1)(N–2)P_2(**B**)–(N–1)(N–2)

2(N–1)(N–2)**c**–(N–2)(**dc+ac+bc**) –2(N–2)(**d+a+b**) +2(**da+db+ab**) = N(N–1)(N–2)P_2(**C**) – (N–1)(N–2)

2(N–1)(N–2)**d**– (N–2)(**ad+bd+cd**) –2(N–2)(**a+b+c**) +2(**ab+ac+bc**) = N(N–1)(N–2)P_2(**D**) – (N–1)(N–2)

We see that these N equations are not independent, as the right side:

N(N-1)(N-2)(P_2(A)+P_2(B)+P_2(C)+P_2(D)) – N(N-1)(N-2) = 0 is verified if P_2(A)+P_2(B)+P_2(C)+P_2(D)=1 *for any N.*

The coefficients of **a, b, c, d** are $2(N-1)(N-2) -2(N-2)(N-1)$ in the sum of the left side are 0

The coefficients of **ba, ca, da, cb, db, dc** are $-2(N-2) + 2(N-2)$ in the of the left side sum are $= 0$

(easy to see as a term as **bc** appears twice in the N–2 equations corresponding to $P_2(\mathbf{A})$ and $P_2(\mathbf{D})$)

In the case p=2 (maximum = 3) the system of equations is also not independent to determine a, b, c, d.

We check that with a=1; b=c=0 and N=3, we can get $P_2(\mathbf{A}) = 1$ this time with p=2.

Finding a,b, c,d for N=3 and given values of $P_2(\mathbf{A})$, $P_2(\mathbf{B})$, $P_2(\mathbf{C})$, $P_2(\mathbf{D})$ is a "quadratic programming problem", and a "pth order polynomial programming problem" for a "(p+1) maximum number of UL policy" in general.

12.2.5 Finding a, b, c, d... Given P(A),P(B),P(C),P(D): Implementation with the Monte-Carlo or the "Simulated Annealing" Method

As it is quite complicated and not justified to use non linear programming for small values of *N*, we use the Monte-Carlo method to resolve the problem. It will work also for the case *p*=1 (which could be resolved as a linear programming problem not too difficult).

Choose randomly, 0<= a,b,c,d < =1 ,then compute P(A), P(B), P(C), P(D). Repeat the choice a large number of times until P(A), P(B), P(C), P(D) are all close enough to the sought values. Of course, for certain values of P, it will not be possible (for example, N=3, p= 1, P(A) = 0.7 is not possible!).Note that we use a better function than the "rand" of the C library which does not give a good uniform distribution for a draw in [0,1].This brute force method can be improved in terms of convergence by the more elaborate "simulated annealing" method. See the 4 examples of Table 12.1 with 4 sets of target probabilities P(A), P(B), P(C), P(D). The columns a, b, c, d gives the solutions a, b, c, d which are used for the selection rules when the UL arrive from the 4 network in order to achieve the best approximation of the target probabilities.

Result for N=4, p=2, P(A)=0.38, P(B)=0.32, P(C)=0.2, P(D)=0.1, with 1000000 Monte-Carlo draws that a, b, c, d do not have a unique solution, so that each problem solution is different while giving the same close approximations for P(A), P(B), P(C), P(D)

Table 12.1 Solutions obtained by a Monte-Carlo approximation

a	b	c	d	P(A)	P(B)	P(C)	P(D)
0.930	0.823	0.553	0.291	0.379	0.323	0.198	0.100
0.800	0.678	0.446	0.208	0.384	0.316	0.200	0.100
0.921	0.808	0.554	0.280	0.380	0.320	0.202	0.098
0.759	0.657	0.425	0.187	0.378	0.320	0.202	0.100
0.551	0.455	0.242	0.034	0.381	0.321	0.201	0.097

Table 12.2 Optimal solution with Monte-Carlo for a trivial case

a	b	c	d	P(A)	P(B)	P(C)	P(D)
0.625	0.525	0.308	**0.100**	0.379	0.319	0.200	0.102

In order to have a more intuitive solution, we set arbitrarily one of the $a[i]$ = $P(i)$ corresponding to the smallest P, that is, for $P(D)=0.1$. It gives as shown in Table 12.2.

No $a[i]$ is null, so that there are cases when a first UL from D is drawn first time. No roaming partner will fill to be excluded.

As you can see, the Monte-Carlo is very simple to implement for this problem and gives stable results, with $P(A)$, $P(B)$, $P(C)$, $P(D)$ close to the objective.

12.2.6 Generalization to the Any p < N Case

Define the following functions:

$K0(A) = 1/N$ **0th order polynomial**

$K1(A) = (1/N(N-1))x((1-b)+(1-c)+(1-d)+..+(1-f))$ **1st order polynomial**

$K2(A) = (1/N(N-1)(N-2))x((1-b)(1-c)+(1-b)(1-d)+..+(1-b)(1-f)+(1-c)(1-b)+ (1-c)(1-d)+(1-d)(1-b)+(1-d)(1-c)+..+(1-d)(1-f)+...)$ **2nd order polynomial**

$K3(A) = (1/N(N-1)(N-2)(N-3))x((1-b)(1-c)(1-d)+..+(1-b)(1-c)(1-f)+(1-b)(1-d)(1-c)+..)$ **3rd order polynomial**

...

$Kp(A) = (1/N(N-1)(N-2)(N-3)..(N-p))x((1-b)(1-c)(1-d)...(1-e).+..)$ **p^{th} order polynomial**

We can then define a recursive computation for Pp(A), Pp(B), ..,Pp(F):
p(A)= (K0(A) + K1(A) +..+ Kp-1(A)) x a + Kp(A)
for the "SoR by SS7 system", which is then very simple as there is no need
of "context" or memory of the past SoR cases to guarantee the desired
distribution of the roaming partner's networks usage.

12.2.7 Case of a non-uniform distribution of the UIs

In reality due to many reasons: incomplete coverage, some operator
installing a high density of base stations at the landing points such as
airports the distribution of ULs is far from being uniform. It is possible to
estimate this distribution using the statistics of the 1rst UL. Let $P_0(A)$,
$P_0(B)$, $P_0(C)$, different of 1/N as in 12.2.2, be the observed distribution of
the 1rst UL. How is the optimization of a, b, c .. modified?
The equation (12.2) on page 216, for p= 1 (at most one reject) which was
for a uniform distribution assumption becomes.
$P_1(A)=P_0(A)$ x a + $P_0(B)$ x(1 -b) x ($P_0(A)((P_0(A) + P_0(C)))$) + $P_0(C)$ x(1 -c)
x ($P_0(A)/((P_0(A)+P_0(B)))$)

Application, case: N=2
One obtains
$\qquad$ $P_1(A)=P_0(A)$ x a + $P_0(B)$ x (1 -b) x ($P_0(A)((P_0(A))$) .
If we take 1-b = a and as $P_0(A) + P_0(B) = 1$,
$\qquad$ $P_1(A) = a$, and same for B:
$\qquad$ $P_1(B) = b$
The result was intuitive in this simple case p=1 and N=2, there is no need
of the « simulated annealing method ».

12.3   Seamless SS7 Routing Architecture to Insert a GLR in the International Signaling Message Flow

In Figure 12.1, we see the *GLR inserted physically between the GMSC and
the IGP*. A much better approach, which is logically the same, is described
in Figure 12.2, *the insertion is logical and only for the SCCP signaling*.
Physically, the international connection is still between the IGP and the
GMSC but all the signaling messages from and to the IGP go through the
GLR. The main advantage is that the *GLR does not have to switch some
TDM voice circuits or the ISUP signaling*. A second one is that all the
international links handling is not changed (in particular, parameter setup

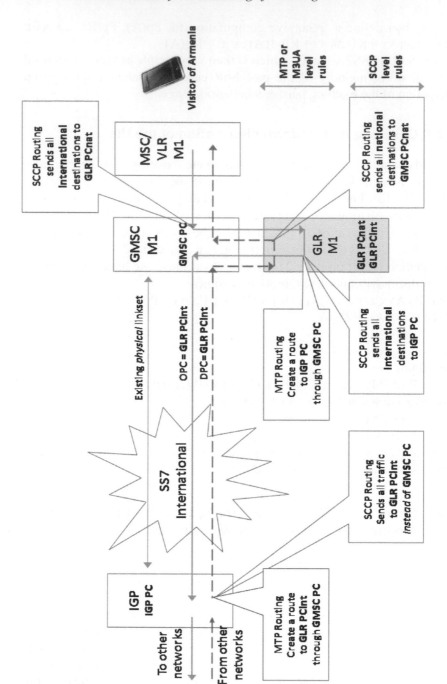

Figure 12.2 "GLR insertion" architecture with a second international Point Code.

for long delay satellite links), the GMSC still handles the MTP2 or IP transmission level.

A most important advantage is also that this architecture where the GLR is logically inserted between the IGP and the GMSC *does not require any "Calling Party-based routing" in the GMSC* (these features are not available in many vendors software, only in dedicated STPs or are optional charged features).

The GLR becomes the new logical international SCCP gateway of the MNO.

The GLR has a national Point Code (GLR PCnat) and is also assigned a second new international point code GLR PCint) (International type; if the MNO is connected to the IGP with a National Point code, GLR PCnat is used). For the explanation, we assume this GLR PCint case.

The IGP is asked to route all SS7 signaling to the new international point code using the *links with the current international point code.* In the example below, we assume the GLR is installed in M1 (+65968 Singapore) as in Figure 12.1. For a smooth cutover, there is an intermediate testing step 3):

1. In the GMSC a route (MTP3 or M3UA) is opened between with the two GLR PCnat and GLR PCint of the GLR.
2. Configure first the two final MTP routing rules of Figure 12.2 in the GLR and in the IGP. From the GLR, it is possible at this stage to test that the IGP Point Code is reachable (destination PC IGP available). The traffic flow is not changed.
3. Choose *one testing destination partner* such a +37493(Armenia) and set in the M1 GLR temporarily the Alias GT method 2 (see [0.8] Figure 2.3) HI+OI_hplmn+NNI_msc, just for this partner (the sent UPDATE_LOCATION SCCP messages concerning Armenia visitors will have +65968 37493 XXX as SCCP CgPa). Create in the MSCs M1 a routing rule for all messages to +37493 to be routed to the PC GLRnat. At this stage, the UPDATE LOCATION process of the Armenia visitors will work in M1 through the GLR. For all the other partners, the flow of messages is not changed.
4. When it works, activate *simultaneously* the full SCCP routing (all the roaming partners) in the IGP and in the GMSC M1, after suppressing the partial routing rule 3) in the GLR.
   The GLR is now used by the M1 MSCs as the logical IGP for all outgoing traffic, and the IGP uses the GLR as the logical GMSC of M1.

The reason for this test plan is that the IGP is not able, in general, to implement a dedicated routing. In the plan at the testing step (3), the routing of responses from the IGP to the GLR is performed by a temporary Alias GT routing scheme set in the GLR.

Solution of Eiffel's question (use of the plane transformations group theory)

a) A point rotation $R(O_1, \omega_1)$ of M around O_1 with angle ω_1 is equivalent to the product of 2 line symmetries with respect to any line D through O_1 and another line D' through O_1 having an angle $\omega_1/2$ with D. If these symmetries are noted S_D and $S_{D'}$ we can write $R(O_1, \omega_1) = S_D \times S_{D'}$

b) The product of 2-line symmetries through the same line D' is the identity, that is, $S_{D'} \times S_{D'} = I$

c) For the second rotation $R(O_2, \omega_2)$, we take the first line symmetry to be around the previous line D', the second being around D'' with D' and D'' having an angle $\omega_2/2$. $R(O_2, \omega_2) = S_{D'} \times S_{D''}$

d) Combining a), b), c) we have $R(O_1, \omega_1) \times R(O_2, \omega_2) = S_D \times S_{D'} \times S_{D'} \times S_{D''} = S_D \times I \times S_{D''} = S_D \times S_{D''}$. It is a point rotation as the product of 2 line symmetries.

e) As D' must be through O1 and O2 according to c), D' is then O_1O_2. D has an angle of $\omega_1/2$ with O_1O_2, D'' has an angle of $\omega_2/2$ with O_1O_2, they intersect at Ω which is then the center of the product of the 2 rotations. The angle $O_1\Omega O_2$ is $\Pi - (\omega_1/2 + \omega_2/2)$. and the angle of the product rotation is the double that is $2\Pi - 2(\omega_1/2 + \omega_2/2)$, that is, a rotation of total angle $-(\omega_1+\omega_2)$ around Ω.

13

Location Services (LCS) in 4G and 3G synchronized Radio Access Networks

Wherever I wonder, wherever I rove,
My heart is in the Highlands wherever I go,
    −Robert Burns (1759–1796)

Most operators are opening 4G networks closely linked with their 3G network. The 3G is used for voice calls and SMS (this is called Circuit Fall Back (CFB)). The reasons presented are:

- Voice quality VoLTE insufficient (this is steadily improving),
- Limited LTE coverage, and the handover 4G <-> 3G is not working well enough during a call as it requires a full IMS support implementation,
- IMS not widely available (one of the architectures allowing to receive calls and SMS over IP).

Note that the CDMA networks cannot use CFB as they do not have the equivalent of 3G UMTS networks. They have, thus, to jump to VoLTE and work on the quality of VoIP in this LTE environment. They raise another reason which is false, that is the inability of the LTE network to provide an adequate precision for the emergency services.

13.1   Location Services Architecture for LTE

An e-SMLC [13.1] is very closely derived from a 3G SMLC using DIAMETER-based primitives directly derived from their MAP counterparts (Table ?.? in Chapter ?), it is not a major effort to implement one from a 3G SMLC with all the positioning methods:

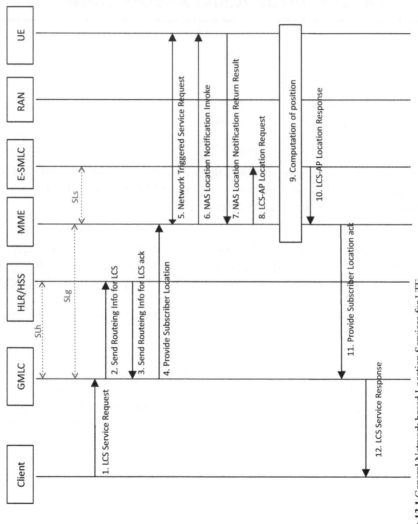

Figure 13.1 General Network based Location Services for LTE.

- Cell ID and E-CID (Cell ID with the use of Timing Advance or power measurements), this is a circular method (intersection of circles),
- U-TDOA (uses "Location Measurement Units(LMU) at known precise locations), this is the "Uplink" hyperbolic method widely installed in the USA to meet the FCC requirements,
- A-GPS (Assisted GPS),
- E-OTD ("Downlink" hyperbolic method).

The SEND ROUTING INFO FOR LCS returns also the MME name, the H-GMLC IP address; AAA server IP address, the PPR (Private Profile Register) IP address which may be used for additional privacy services [1.5] and [1.6].

Comparing with [0.9 Chapter 6], as the RNC or BSC function is integrated in the MME, there is no difference for the 4G LCS architecture between "NSS-based" and "BSS-Based", as the Location Request (8) is always initiated by the MME.

The SLh and SLg protocols are based on DIAMETER/SCTP (port 9082). The MME creates a permanent SCTP association with the e-SMLC when initialized. This association is also used by the e-SMLC to carry transparent messages for:

LPPa protocol with the eNOdeBs,

LPP Protocol with the UE, (not used in this chapter, it is used for A-GPS),

while the SLmAP protocol is direct between the eSMSC and the LMUs if used.

13.2   Trace on the GMLC<->MME SLg Interface

It is exactly the same as on the interface GMLC<->MSC for GERAN an UTRAN Radio Access Networks below, except that the protocol Lg/DIAMETER is used. The parameters are the same as well as the coding of the position and the Positioning Data which shows which method is used.

13.2.1 Provide Subscriber Location Request

```
- - - - Super Detailed SS7 Analyser (C)HALYS - - - - - - -
    PA_Len = 34
  PROVIDE-SUBS-LOCATION-REQ(61)
    timeout(45)
```

```
       L = 002
       Data: timeout value =  36 sec
    invoke_id(14)
       L = 001
       Data: 1
    Location_estimate_Type(144)
       L = 001
       Data: (0):current location
    LCS_Client_Type(145)
       L = 001
       Data: (0):emergency Services
    mlc_number(137)
       L = 007
       Data: Ext = No extension
          Ton = International
          Npi = ISDN
          Address = 14044XXXXXX    /* USA */
    imsi(18)
       L = 008
       Data: Address = 310260572YYYYYY  /* USA */
- - - - - - - - - - - - - - - -- - - - - - - - - - - - - - -
```

13.2.2 Provide Subscriber Location Ack : Mobile Switched Off: Poor Accuracy , the Cell ID Method is Used

```
- - - - Super Detailed SS7 Analyser (C)HALYS - - - - - - -
   PA_Len = 37
   PROVIDE-SUBS-LOCATION-CNF(184)
    invoke_id(14)
       L = 001
       Data: 1
    Location_estimate(3GPP 23.032)(153)
       L = 008
       Data:
          23.032 location coding
            Type of Shape: (1)Ellipsoid point with uncertainty circle
             latitude-longitude = 28.3112N 81.4771W
             uncertainty in meters = 1410.4  /* poor accuracy: Cell ID method used */

    age_of_loc_estimate(158)
       L = 001
       Data: 0
    UTRAN_Positioning_Data(TS 25.413)(633)
       L = 003
       Data: (Hex) 400C60
          discriminator: (0):usage of the non GANSS(Galileo and Additional Navigation
Satellites) methods
          list of methods used
```

    method: (1):Reserved(1)
    usage: (4):Attempted successfully: case where MS supports multiple mobile
based positioning method and the actual method used by the MS cannot be determined
    method: (12):Cell ID
    usage: (3):Attempted successfully and used to provide location

  cell_ID(49)
   L = 007
   Data: MCC = 310 MNC = 260 LAC = 32699
     Cell_ID = 62611
  SAI_Present(165)
   L = 000
- - - - - - - - - - - - - - - - -- - - - - - - - - - - - - -

13.2.3   Provide Subscriber Location Ack: Mobile Switched on: FCC Standard Accuracy, Accurate Positioning with LMUs

This request is for the same handset in the same area, but this time it is switched on and it can provide the LMU measurements (LMU are installed in that area).

  - - - - Super Detailed SS7 Analyser (C)HALYS - - - - - - -
   PA_Len = 37
   PROVIDE-SUBS-LOCATION-CNF(184)
    invoke_id(14)
     L = 001
     Data: 1
    Location_estimate(3GPP 23.032)(153)
     L = 008
     Data:
         23.032 location coding
           Type of Shape: (1)Ellipsoid point with uncertainty circle
           latitude-longitude = 28.3111N 81.4792W
           uncertainty in meters = 148.6   /* good accuracy: U-TDOA with LMUs
method is used */

    age_of_loc_estimate(158)
     L = 001
     Data: 0
    GERAN_Positioning_Data(TS 25.413)(633)
     L = 003
     Data: (Hex) 0043
         discriminator: (0):usage of the non GANSS(Galileo and Additional Navigation
Satellites) methods
         list of methods used
         method: (8):U-TD0A(Uplink Time Difference of Arrival)

usage: (3):Attempted successfully: results used to generate location

cell_ID(49)
  L = 007
  Data: MCC = 310 MNC = 260 LAC = 33608
    Cell_ID = 26363

The price of standalone LMUs is not high and LMUs are included in eNodeB and the recent 3G BTS: the U-TDOA method is quite attractive as it works even without the GPS being activated in the handset or any LCS software. A basic feature of an LMU (a GPS receiver in most cases) is that they are *time synchronized to a common reference*.

13.3 The Uplink Time Difference of Arrival(U-TDOA) Positioning Method

13.3.1 Difference between U-TDOA and E-OTD

The popular U-TDOA method is applicable in GERAN, UTRAN and eUTRAN.

In [0.9, page 110], the previous 3G- E-OTD method is explained in detail, it is a "downlink" Time of Arrival method as the measurements are made by the MS. It requires handsets which support E-OTD, but could work with Base Stations (BS) which are not time synchronized. Since a few years, the UTRAN BS and of course the eNodeBs are time synchronized (usually with a GPS receiver). This has allowed the U-TDOA method with LMUs installed at important areas in the network. This allows the U-TDOA LCS method (U = 'Uplink') where the measurements are made by the LMUs under the control of the SMLC. It handles then all the legacy phones without a special LCS feature.

It is simpler than E-OTD as no assistance data are needed (the calibration RTDi giving the time difference of each BS # i with respect to a common reference).

Physically, most recent 3G BTS and all eNodeB include the LMU function (LMU B). In the explanation, we will say that the measurements are made by the LMU. The LMUs even if physically integrated in an eNodeB, are a separate entity which communicates directly with their serving e-SMLC using the protocol SLmAP [13.6]

You can see that the LCS call flow of Figure 13.1 for LTE is not different from what was used in GERAN and UTRAN networks. We will use the UTRAN case (3G) and in the last section adapt to the LTE case.

The initial connection of the LMU to the e-SMLC is not shown in the detailed position computation method of Figure 13.3, it is explained below.

13.3.2 Initial Connection of the LMUs to Their Serving e-SMLC and Further Exchanges

The e-SMLC would not need to have a database of the longitude-latitude of the LMUs as the LMUs have an integrated GPS which can provide it when they are inititialized. The *LMUs create a connection to the e-SMLC* using a SCTP socket, and sends a SLmAP SETUP REQUEST which contains:

- LMU information (LMU position latitude-longitude provided by their GPS, List of RF bands),
- LMU ID (unique ID within an eUTRAN).

When the e-SMLC needs to have a LMU # *i* performing a Measurement, it sends a SLmAP MEASUREMENT REQUEST to this LMU # *i on the existing SCTP connection* which contains

- e-SMLC Measurement ID (this is the ID listened by the LMU in the SRS broadcasted by the UE),
- UL RTOA Measurement Configuration (containing the "*SRS format*" specific of a given UE).

When this LMU # *i* receives, during the specified period, a SRS with this e-SMLC provided SRS format, it will return to the e-SMLC a SLm MEASUREMENT RESPONSE which contains the U-TDOA measurement:

- e-SMLC Measurement ID,
- UL RTOA Measurements ($Ti - T$ below).

It is *not the IMSI which is broadcasted* by the UE in the SRS messages where they are triggered by the e-SMLC to broadcast, it is this *SRS format*, which is made common for a measurement for the receiving LMUs (they listen to it) and for the broadcasting by the UE of a specific Sounding Reference Signal (SRS) customized by the selection of a SRS format.

13.3.3 Principle of the U-TDOA Hyperbola Intersection Method

The SMLC receives a request which was generated by a GMLC to provide the location of a customer. From the Cell Id which is contained in the BSSAP-LE Perform Location Request, it can find the LMUs (at least three) close to the customer. The LMUs are addressable by an IMSI and from a table the SMLC knows from which VMSC they are dependent from. The overall procedure is described in more details by Figure 13.3.

1. The SMLC sends to each selected LMU a MESASUREMENT REQUEST asking to listen to an identifiable signal from the UE
2. Unlike in the E-OTD method where a power burst is sent *by the Base stations*, the SMLC triggers the handset to send a particular signal called Sounding Reference Signal (SRS) *which is received by selected neighboring LMUs.*
3. This SRS containing the handset time T is received by the surrounding BS and LMUs. The LMUs which have been selected by the SMLC will return $Ti - T$ (Ti is their reception time) to the SMLC to compute the position.

If the handset and the BS # i time were synchronized, $c \times (T - Ti)$ would give the distance customer-LMU # i (c is the speed of light), and the location could be computed as the intersection of circles.

This is not the case and the difference of transmission time from the handset to BS # i and BS # j is computed by the SMLC as:

$$GTDij = (Ti - T) - (Tj - T),$$

where the unknown time offset of the handset vanishes.

The difference of distances to the two BS is

$$UTDOAij = c \times GTDij$$

and the location is estimated as the point closest to the intersection of the hyperbolas having their foci at the LMUs (see Figure 13.2). In [0.7 Chapter 13], there is the exact mathematical solution using Gauss' resultant theory and Sturm's algebraic polynomial real roots resolution. This solution is the global minimum of the non convex polynomial:

$$\min Q(X,Y) = \Sigma_{ij}(Pij(X,Y))^2, \qquad\qquad\qquad [13.1]$$

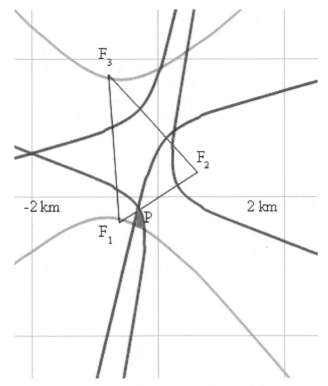

Figure 13.2 Polygonal approximation of the uncertainty of the hyperbola intersections.

where each *Pij(X,Y)* is the quadratic equation of one of the hyperbolas.

$Q(X,Y)$ is degree 4 and has *several local minima* in which zero is the gradient of *Q(X,Y)*. This gives two equations of degree 3 in X and Y. Gauss's resultant theory [0.7 Section 13.7.2] allows to eliminate the variables resulting in a one variable degree 9 polynomial. Sturm's method [0.7 Section 13.8] allows to find upto nine real roots which are a local minimum. *Q(X,Y)* is evaluated at each to give the *global minimum* P.

13.3.4 The Polygonal Uncertainty Shape

[13.1] gives the list of standard uncertainty shapes, circles, ellipsoid, ellipsoid arcs, etc. A polygon shape is also included. It comes from trying to approximate the uncertainty of hyperbolic intersection methods such as E-OTD and U-TDOA.

F1, F2, F3 are the positions of the three LMUs which have been selected for the positioning and are *the foci of three hyperbolas*. There is a shaded area limited by three hyperbolas which could be considered as the "uncertainty zone", P being the solution of [13.1].

Mathematically, many SMLCs are rather crude, and the author has never audited an operational one which provides the polygonal uncertainty shape in the "Positioning Data", as they could for U-TDOA.

To obtain a mathematically coherent approximation of the polygonal uncertainty zone in the general case, is a difficult problem. Here is a heuristic method:

1. approximate the position of the handset using [0.7 chap. 13] shown as P, this is not necessarily an interior point on the nonlinear uncertainty domain.
2. find the real x–y points which are intersections of all the n hyperbolas (up to four real points for each pair). The resultant and Sturm's methods can be used to find all the real solutions of the 2-combinations of the quadratic equations (one for each pair of LMUs):

    $Pij(X,Y) = 0$

3. find the convex polygon which approximates the uncertainty area bordered by one of the hyperbola. For each 2–hyperbola intersection, it is losange type shape.

13.3.5 Performance Measurement of the Resultant-Sturm Method

The computation time depends very little on the number of hyperbolas or the number of circles when power measurements are used instead of LMUs (quadratic Cartesian equations). A formal calculus is made of the polynomial coefficients of the resultant matrix [0.8 Chapter 6]. Then the formal calculus of the determinant and the resolution of its real roots is independent of the number of quadrics; it depends only on the resolution uncertainty for the roots set in the Sturm polynomial root algorithm.

A 2013 Intel processor (6 core, 12 threads) is able to compute 2100 locations/ second. So a very cheap server with two processors can handle over 15 million locations in the peak hour with this accurate method.

13.3.6 The Sounding Reference Signal(SRS)

This is standardized in [13.9]. It is used to test the transmission quality of an Uplink, using the LTE PCCHU radio channels used for signaling. It is an audio sound signal *sent by the UE* with various parameters allowing to create a *unique signal* which can be detected by a signal correlator in the LMUs. To trigger an Uplink Time of Arrival measurement, the eSMLC asks the servicing eNodeB to set the UE with a characteristic SRS format and a repetition period, in order to be listened and detected *by the neighboring* LMUs (9.6 of Figure 13.3).

13.3.7 Protocol Details to Implement an e-SMLC Able to Use LMUs

Figure 13.3 details the box 9 i.e. "Computation of Position" of the overall Figure 13.1.

Explanation:

8.—The e-SMLC receives the LCS-AP Location Request from the MME with the IMSI and the servicing eNodeB of the target UE.

9.1—It sends on the established SCTP connection with the MME a LLPa PDU UTDOA INFORMATION REQUEST to the eNodeB for the UE. It sets the SRS transmission characteristics parameter described in [13.7]. The eNodeB selects an available "SRS format" which allows the UE, within the visited cell to send a unique Uplink SRS signal identifiable by the LMUs which will be selected to listen. The eNodeB sends the SRS format to the UE over the air interface.

9.2—The eNodeB returns this SRS format for the target UE.

9.3—The Cell ID provides the e-SMLC with a crude location of the UE which allows to select the *n* closest LMUs.

9.4, 9.4'—The e-SMLC asks the selected LMUs to listen to the SRS format which is now being sent periodically (9.5) by the UE.

9.6, 9.6'—The LMUs return in the UL RTOA parameter their observed Time of Arrival of the SRS, $Ti - T$.

9. 7—The position of the UE is computed using the method of 13.3.3.

10.—The e-SMLC returns the position of the MME which will format it in the PROVIDE SUBSCRIBER LOCATION Ack returned to the GMLC.

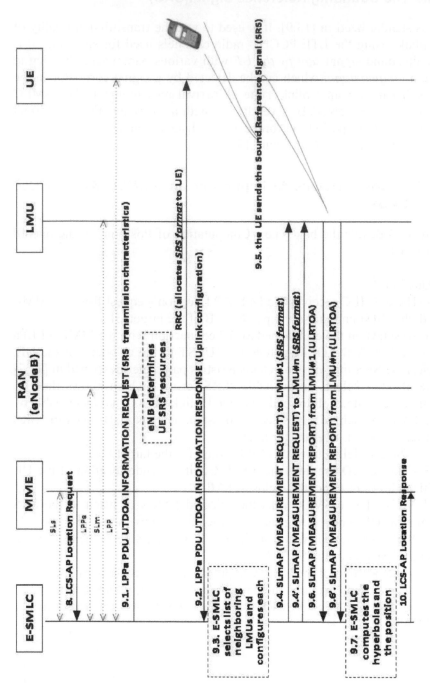

Figure 13.3 Uplink TDOA computation of position procedure with LMUs.

13.3.8 Various Improvements for the U-TDOA and Other Methods

The method explained is quite basic. Because the SRS signal may be weak, it would be advantageous to repeat the signal so that the correlation period of the LMUs may be extended (PRS may be used). Another method is to select a close "reference LMU" and to get a first crude estimate of the position of the UE and the e-SMLC can compute an estimate of the SRS delay to the other LMUs. This way, narrower SRS search parameters can be sent. Some LMUs also integrate a direction finder and can provide additional Angle of Arrival measurements which contribute to the accuracy.

To use methods other than U-TDOA, one could use other LPPa commands such as

E-CID MEASUREMENT INITIATION,
OTDOA INFORMATION REQUEST.

13.4   Mobile Advertisement Service Architecture

The purpose of mobile advertisement projects is to provide alerts when a subscriber gets in the vicinity of a commercial area which rents the service to the MNO, so a commercial message may be sent by SMS, push WAP, etc. In the 3G networks, there is a feature called Area Event Report (see Figure 13.4) which was designed for this purpose, but we will see that this would meet the objective very partially, and another architecture which monitors the area changes at the BTS or eNodeB level is required.

13.4.1 Use of Triggered Change of Area for 3G: limitations

The change of area Location Request is started with a PSL "Deferred" (13.2.1 is the case of an Immediate Request) which defines the area (three cells) of the commercial centre:

```
- - - - Super Detailed SS7 Analyser (C)HALYS - - - - - - -
   PA_Len = 34
   PROVIDE-SUBS-LOCATION-REQ(61)
     timeout(45)
      L = 002
      Data: timeout value =  36 sec
     invoke_id(14)
      L - 001
      Data: 1
     Location_estimate_Type(144)
```

```
       L = 001
       Data: (3):activate deferred  location
     Deferred_Location_Event_Type
       L = 001
       Data: (1):entering into area
     Area_Type
       L = 001
       Data: (5): UTRAN cell type
     Area_identification
       L = 007
         310.260.33608.26363
     Area_Type
       L = 001
       Data: (5): UTRAN cell type
     Area_identification
       L = 007
         310.260.33608.1203
     Area_Type
       L = 001
       Data: (5): UTRAN cell type
     Area_identification
       L = 007
         310.260.33608.1204
     LCS_Client_Type(145)
       L = 001
       Data: (0):emergency Services
     mlc_number(137)
       L = 007
       Data: Ext = No extension
          Ton = International
          Npi = ISDN
          Address = 14044XXXXXX      /* USA */
     imsi(18)
       L = 008
       Data: Address = 310260572YYYYYY  /* USA */
```

When the event occurs, the call flow is given by Figure 13.4.

The Location Request (Deferred type) for change of area and periodic alerting is:

-    only for a given subscriber,
-    only for a small list of areas (UTRAN Cell ID of eUTRAN Routing Area) limited to 10 by the PROVIDE SUBSCRIBER LOCATION parameters.

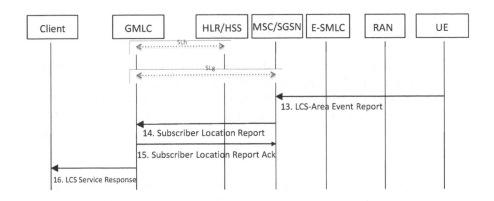

Figure 13.4 Deferred MT Location Request for change of Area.

In Figure 13.4:

13.—the UE sends a LCS-Area Event Report asynchronously when the change of area conditions sent in the initialization PROVIDE SUBCRIBER LOCATION are met (not shown in the figure).

14.—the MSC or the SGSN sends a MAP Subscriber Location Report to the requesting GMLC.

The LOCATION REPORT uses the same format as in the immediate PROVIDE SUBSCRIBER LOCATION ack.

15.—A LCS Service Response id sent to the client.

If the client application wants then an accurate position (not just the cell ID), it sends an immediate location request (not shown).

In this architecture, the subscribers or visitors must be selected and the list of areas sent to their UE. The alert conditions are flexible: entering the area, being in, leaving as well as the interval between alerts if they meet the alert condition again. This is not quite the mobile advertisement requirement, to send the publicity message only once.

This is not applicable in LTE as the Deferred Location Requests are not available in eUTRAN for some reason which are ignored by the author.

13.4.2 Efficient Mobile Advertisement Architecture Possible in LTE

13.4.2.1 Blind Areas for Handovers in 3G

Another architecture is then required (rather simple to provide in the LTE networks). In 2G, 3G and LTE there are two cases when a change of cell in the Circuit Domain or the Packet Domain (same principle) creates signaling messages which can be monitored:

Idle case: when the handset is idle and changes the LAC(CS of 3G) or RAC(PS) area. He finds that the signal is not sufficient in the current area and registers (UPDATE_LOCATION or UPDATE_LOCATION_GPRS) in the MSC/VLR or SGSN covering the new selected area using TMSI or IMSI in the case of a new MSC or SGSN. This is necessary so he can be paged to receive a call or SMS at any time, the MSC/VLR must know the current area.

Active case (Handover with a change of cell): For the Mobile Advertisement application, one needs to be alerted by an alerting message when an *idle handset* gets in the vicinity of a selected commercial Cell ID. In 3G, this can happen only when this cell is at the border of two LAC or RAC, the other cells are in "blind areas" where the service cannot be provided.

In GERAN or UTRAN there are *periodical changes of location* (a parameter broadcasted on the radio control channels gives the period from 6 minutes to 24 hour) which are *decided by the handset* which selects a new Cell. It selects the new cell and *the **new base station** informs* the VLR trough the BSC or RNC with a BSSMAP LOCATION UPDATING REQUEST containing the TMSI and the new selected cell. The handset stores this new current cell ID in one of the SIM card files called Eflocinfo (this is used for Handover).

If the new cell is in the same MSC, the VLR does not trigger a MAP LOCATION UPDATE to inform the HLR. This is why if a SMLC is periodically interrogating a VLR for the last visited cell, even if the handset is still, one sees changes of the last visited cell. *The SMLC which would be polling regularly the VLR cannot know if it corresponds to a real displacement of the handset.*

There is also the changes of location for *the handover case* (the handset is having an active call which is continued from one cell to the other or even from one VLR to another one (inter VLR handover). In that case also the VLR is informed, but *it does not mean that any alert message is generated toward some mobile advertisement SMLC*. Contrary to the idle periodical change of location, the change is decided by the BSC or RNC based on the power measurement reports received from the handset. The controller sends a RR HANDOVER COMMAND to the handset through the *old base station.*

In none of the 2 cases periodical changes of location or handover are there any possibility of alert messages, except installing probes on all the BSC or RNC links to the VLR to monitor the BSSMAP LOCATION UPDATING REQUESTs.

There are also perodical changes of location and handovers in 4G eUTRAN. The major difference is that the protocol between the MME and the PGW allows the transmission of alert messages (Modify Bearer Request) with a new ULI (Cell ID) whenever there is cell change due to any of the two reasons. For this, see Figure 13.5, the PGW should have set the parameter "inform change of location" in the Create Session Response. It is easy then as there is a single equipement, the PGW, where alerts arrive, to have the PGW sending an alert message to an external marketing application.

13.4.2.2 Handover Call Flow in LTE

There are two main types of handovers:

- Intra-eUTRAN (the UE remains in the same LTE network)
      -with the X2 interface,
      -with the S1 interface (this will be always the case if there is a change of MME (inter-MME handover)) .
- Inter-RAT (the UE moves between UTRAN and eUTRAN): We leave this aside and refer to [13.11] for a full description of the Handover procedures in the various cases. The purpose of this section is to identify the simplest monitoring point in LTE to perform Mobile Advertisement.

In eUTRAN, *the Hand Over decision* (full description of the procedures in [13.11]) is taken *by the source eNodeB* which has measurements reports *of both the source and the target eNodeB*; this is handled by the RNCs in UTRAN.

The S1 based handover of Figure 13.5 (simplified) is not much different from the X2 case, as it is more general, we use the case. If X2 is available, the messages between the source and target eNodeB are through direct links. The handover architecture allows to switch cells smoothly without interrupting the established user packet data flow.

Handover preparation
1.   The UE send regularly Measurement Reports to his (serving) Source eNodeB,
2.   The Source eNodeB makes a decision to switch to a new Target eNodeB,

3. A Handover Request [13.10] is sent by the Source eNodeB to the Target eNodeB and acked.

Handover execution
4. It starts with a Down Link allocation sent to the UE then,
5. An RRC Connection Reconfiguration request is sent to the UE,
6. The Uplink data packet is sent to the Target eNodeB,
7. The Source eNodeB sends an eNodeB Status Message to the target to convey the status,
8. At this stage, the Target eNodeB buffers the user data packets,
9. Synchonization between the UE and the Target eNodeB and Uplink allocation confirmation sent from the target eNodeB to the UE giving the new value of the Timing Advance.
10. The RRC Connection Reconfiguration is received from the UE and the Uplink user data packets are sent out.

Handover completion
11. The Target eNodeB asks the MME to switch the down link path to itself,
12. To perform this action, the MME sends a GTP Modify Bearer Request to the S-GW. It contains (case of Mobile Marketing service) the User Location Info (new UTRAN cell ID) if it was requested by the PGW in the Create Session Response for this UE. This GTP Modify Bearer Request is relayed to the PGW which can note the cell ID change,
13. The SGW switch the Downlink path and the uplink and downlink user data flow is resumed,
14. The GTP Modify Bearer Response is sent to the MME,
15. The MME answers Path Switch Request Ack to the Target eNodeB,
16. The Target eNodeB sends a UE Context Release message to the source eNodeB.

13.4.2.3 Detecting the Change of Locations in the PDN Gateway

One sees that when the UE has successfully accessed the target cell, the MME sends a MOFIFY BEARER REQUEST to the SGW which relays it to the PDN Gateway. This GTP-C message contains the ME Identity and, if the eNodeB location information change reporting was activated in the GTP Create Session Response by the PDN Gateway, it contains the eUTRAN Cell ID of the new cell. If the MMEs properly implement the standards, *a PDN Gateway can then be a convenient central monitoring point for all the area changes* to implement mobile advertisement.

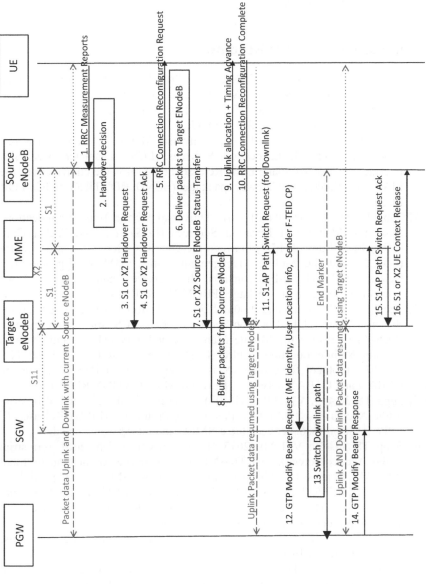

Figure 13.5 Intra-eUTRAN and intra-MME S1 based handover .

In a proprietary implementation of the PDN Gateway, it could trigger the LCS Client Responses (16) of Figure 13.4. Section 3.7.3.2 shows an example of a PDN Gateway which GTP Create Session Response asks the location information change.

References and Further Reading

[13.1]   3gpp   TS   23.271,   V11.2.0   (2013-11),   "Universal   Mobile Telecommunication System(UMTS); LTE; Functional stage 2, description of Location Services(LCS), (Release 11)", the basic document which explains the various positioning methods.

[13.2] 3gpp TS 29.173, V11.1.0 (2013-01), "Universal Mobile Telecommunications System (UMTS); LTE;Location Services (LCS); Diameter-based SLh interface for Control Plane LCS, Release 11". SLh is the equivalent to the MAP LocationServiceGateway package (SEND ROUTING INFO FOR LCS).

[13.3] 3gpp TS 29.172, V11.0.0 (2012-10), "Universal Mobile Telecommunications System (UMTS); LTE;Location Services (LCS); Evolved Packet Core (EPC) LCS Protocol (ELP) between the Gateway Mobile Location Centre (GMLC) and the Mobile Management Entity (MME); SLg interface, Release 11". The SLg contains the services equivalent to MAP LocationServiceEnquiry package (PROVIDE SUBSCRIBER LOCATION).

[13.4] 3gpp TS 29.171, V11.3.0 (2013-06), "Universal Mobile Telecommunications System (UMTS); LTE;Location Services (LCS); LCS Application Protocol (LCS-AP) between the Mobile Management Entity (MME) and Evolved Serving Mobile Location Centre (E-SMLC);SLs interface, Release 11". The SLs protocol for e-UTRAN is the equivalent of BSSAP-LE in the GERAN and UTRAN networks (PERFORM LOCATION REQUEST)

[13.5] 3gpp TS 36.455, V11.3.0 (2013-07), "LTE;Evolved Universal Terrestrial Radio Access (E-UTRA); LTE Positioning Protocol A (LPPa), Release 11". LPPa is the protocol between the e-SMLC and the servicing eNodeB.

[13.6] 3gpp TS 36.459, V11.2.0 (2013-2), "LTE;Evolved Universal Terrestrial Radio Access Network (E-UTRAN); SLm interface Application Protocol (SLmAP), Release 11". The protocol between the eSMLC and the LMUs.

[13.7] 3gpp TS 36.355, V11.3.0 (2013-07), "LTE;Evolved Universal Terrestrial Radio Access (E-UTRA); LTE Positioning Protocol (LPP), Release 11", The LPP protocol between an e-SMLC and an UE.

[13.8] Benedetti, R., and J.J. Rosler, Real Algebraic and semi-Algebraic sets, France: Hermann, 1990.

[13.9] 3gpp TS 36.211, V11.3.0 (2013-07), "LTE; Evolved Universal Terrestrial Radio Access (E-UTRA); Physical channels and modulation, Release 11".

[13.10] 3gpp TS 36.413, V 11.4.0 (2013-07),"LTE; Evolved Universal Terrestrial Radio Access Network (E-UTRAN); S1 Application Protocol (S1AP), Release 11".

[13.11] 3gpp TS 36.300, V11.6.0 (2013-07), "LTE;Evolved Universal Terrestrial Radio Access (E-UTRA) and Evolved Universal Terrestrial Radio Access Network (E-UTRAN); Overall description;Stage 2, Release 11"

14

Advances in Policy Charging an Control (PCC): handling of pre-emptive priorities with the knapsack problem in Advanced LTE

Monica: No, no, seriously, what is it? Is there something fundamentally un-marriable about me?
Chandler: Uh... uh..
Monica: Well?
Chandler: Dear God! This parachute is a knapsack!

−Friends TV Serie, Episode 1.23

14.1 The Need to Implement Pre-Emptive Priorities for 3G or 4G Data Services

This chapter concerns the 3G and 4G data services and recent advanced PCC functions in the GGSN or PGW 4G [14.3]. The need for this design emerges from the situation *that many users can be competing to obtain the radio bandwidth* corresponding to the QoS guaranteed by their subscription. It is also useful to think really of the meaning "priority" as used in queuing theory, does the new entrant wait to have some user leaving the "service station" or does he pre-empts the other users being served currently?

Differing from the classical GGSN or PDN gateway designs, even when equipped with a PCC, it allows to make room for priority users when the visited radio cell (the most frequent case) or the internet connection is used near congestion. This is quite necessary when the infrastructures of a data mobile network are shared between low priority users without really a QoS guarantee, and official security forces or VIP-High priority (considering "VIP" users as users who are ready to pay for higher QoS) . When needed, capacity must be released instantly, more particularly for the *"group call application"* when firemen, or police officers in a crisis situation must be able to talk and share videos (RCS

(Chapter 5) would be useful for this service) showing and commenting the action scene with the required speed and quality.

This design provides pre-emptive priorities for these officials that are when one of their members connects, he will get the *full guaranteed bandwith*, not a fraction, hence the use of *integer variables* below in the mathematical optimization. To achieve this, *lower priority users may be completely offloaded* that is deleted from the radio cell.

If security forces have a top priority QoS from their MNOs, it is also of interest that they *have their own GGSN or PDN gateway* implementing the scheme and prioritizing among their own hierarchical levels.

14.2   Current Implementation of the PCC

The current GGSN and PGW(includes the PCEF) [14.1] implement the PCC function (Policy Charging and Control (the SDP is part of it)) which allows to define individual QoS for the users. Three main components are defined in this architecture: GGSN or PGW, PCEF (with DPI (Deep Packet Inspection) and PCRF (Policy Charging en Regulation Function). The PCRF includes a SDP (Service Data Point) which is a Data Base where the users QoS are recorded, including priorities, bandwidths with various levels. The PCEF may be included in the same system as the PGW or separate.

According to this standard architecture, the PGW dialogs with the PCRF using the Gx protocol based on DIAMETER and communicates with the Internet services using the Gi or SGi physical interface.

Current PCC implementations (that is the set MME + PGW - PCEF + PCRF) are quite crude or void in terms of pre-emptive priority management which is necessary in particular for security forces, as they allocate a new demand in a cell ((note that cell identifier is not a mandatory Gx parameter):

- without having or using the cell radio bandwidth usage.
- And above all without the offload possibility allowing to authoritatively lower a bandwidth or even drop non priority users in the same cell as the priority new entrant,
- It would be commercially advantageous to allow a smoother cohabitation between riches or important users and poor ones. Unfortunately the smooth change (forced HANDOVER) to a neighbor cell under the control of the PCC does not exist nor is supported by the latest standards for MMEs [3.8].

Another difficulty is the amount of packet processing to implement the DPI function of a PCEF. The DPI is necessary to recognize the user traffic: VoIP, download streaming which are large bandwidth consumers without bringing much revenues to the MNO.

This DPI is very demanding in terms of packet processing (example: 90 Gbps for an Asian MNO with 6 million subscribers).

The SON (Self Organizing Network) [14.4] concept is included in the eNodeBs of the 4G radio networks, and in advanced recent 3G UTRAN. This allows neighbouring eNodeB to exchange congestion informations (X2 interface), number of users, etc. in order to transfer the users from a cell to another with eventually a change in priority. An implementation option is C-SON (Centralized SON) with centralized policy decisions. However this does not meet the need of this chapter, a*s the value of the different user categories, especially for the priority users is not accounted for.* In the pre-emptive priority implementation, the global decision is taken when there is a new entrant in a cell, and not dynamically with an anonymous traffic as a SON does.

In a 4G radio network, the data traffic control is provided by the eNodeBs, depending on the QoS they assign the number of Time Slots to each user for the Downlink direction and the Uplink direction. The mobiles only have a limited queuing capability for the UL direction using the capacity allocated by the number of UL Time Slots. In the Downlink direction, the eNodeBs have also a limited queuing capability, the short term regulation must be handled by the PDN Gateway. With pre-emptive priorities the Qos are centrally computed based on the cell capacity usage, and the value of each user's traffic.

14.3 Principle of a PGW with Pre-emptive Priorities Handling

14.3.1 Allocation of Priorities Inside a Cell

The PCC describes below implements a type of C-SON which considers the value of the traffic of the different user classes, while using the standard 3G and 4G GTP protocols for data if no centralized forced HANDOVER is used. So it can interoperate with any other standard MMEs.

For each new Session Request, this PCC will optimize a linear function based on the economic utility of the new entrant and of those already attached on the same cell. A linear inequality models the maximum bandwidth of this cell. Another linear inequality models the maximum bandwidth of the Gi or SGi interface. The *variables are integer*. In 14.3.4 we show that one of the two inequalities dominates the other.

For each Create PDP Context or Create Session Request, it is a classical "knapsack problem" with typically a maximum of 20 variables (small then). Several thousands of problem resolution per second; that is sessions per second are possible with current servers using the Bellmann algorithm.

The PGW interrogates the PCRF (see Figure 14.1) which returns the QoS of the new entrant (the profile is in the Data Base SDP). The cell information ULI known by the PGW is used by the PGW to find all the users in the new entrant's cell and recompute the number of each users with a given priority class. The PGW performs the pre-emptive priority changes by allocating the desired priority and bandwidth of the new entrant, dropping users which cannot any more have their QoS guaranteed, eventually switching them with full service to a neighbor cell with a « forced HANDOVER »eventually (this is not the purpose), reducing the bandwidth of the lower priority users .

The PGW uses it data base BD URR which includes the bandwidth of all the cells based on the number of transmitters and the technology. The PCRF and Gx interface of this design remain standard; the algorithm is implemented in the PGW which also has the GTP commands to modify the QoS of the various users in the concerned cell (the cell could be a SAI (Service Area Identifier) or an eCell ID 4G, it does not change the principle.

To simplify the eNodeB is not shown, only the logical connection between the UE and the MME. Initially, the subscriber profiles are off-line created on the SDP (Service Data Point) which copies the individual QoS profiles to the PCRF. The SDP copies also these QoS to the HSS for each APN (Internet, MMS service, Blackberry, etc.).

In the attach phase 0), the MME will send an UPDATE_LOCATION_Request(GPRS) to the HSS which includes the IMSI and the APN requested for this session. The HSS will return the « QoS allocated by HSS » which is the same as « QoS allocated by SDP ». Both in 4G and 3G (same name), this QoS is sent in an INSERT_SUBSCRIBER_DATA_Request message. The parameter name is called « QoS-subscribed » in MAP [1.5] and S6a [2.1] and is coded according to [14.5]. One has then

QoS APN allocated by HSS = QoS-subscribed

14.3.2 Practical Allocation Numerical Example with the Knapsack Problem

Let 6 policy classes with guaranteed bandwidth (Mbps) a_i be :
20,10,5,1,3,8
Let the 6 economic utility of each policy allocated i c_i be:
80,20,3,1,5,10
Let the capacity of cell#j B_j be :   185 Mbps
Let the number of users of each class d_i be :   2,3,10,30,8,3
    max $P = \Sigma\, c_i\, x_i$, $0 <= i <= N$ number of classes of different   QoS policies
    $\Sigma\, a_i\, x_i <= B_j$,   for the visited cell j
    $0 <= x_i <= d_i$, x_i integer
One finds $x_i = 2,3,7,30,8,3$ with the Bellmann classical algorithm [14.2].

The used capacity at optimum is only 183 (no congestion) when there is a demand for a class #1. Only 7 class #3 policies could be allocated out of a demand for 10 class #3. All the capacity could not be used because the PGW guarantees the requested bandwidth *if the service is granted (no fractional values).*

    With a non standard MME (there may be changes in the standards evolution to include it), the PGW could « offload » certain users on neighboring cells with a forced HANDOVER in order to maintain their service when priority users need bandwidth on their current cell (6 " of Figure 14.2). To perform this we suggest a new GTP command PGW -> MME « GTP Forced Handover » including the current TEID and the new cell identifier on the same MME (or SGSN in 3G). However a policy more standard with existing MMEs could be that the PGW degrades the service quality by activating a secondary PDP context. This offers the advantage that the APN and the terminal IP remain the same so that the session is continued. This is possible because the UEs are evolving to use multiple PDP contexts depending upon which application is activated.

14.3.3 Detailed Implementation Description

Figure 14.1 represents a PGW 4G with pre-emptive priorities.
    The case of a GGSN 3G is identical:

- SGSN replaces MME,
- GGSN replaces PGW,

- Create PDP Context (GTPv1) replaces Create Session Request (GTPv2).

The attach of the User Equipment(UE) to the MME sends an UPDATE LOCATION to the HSS (UPDATE LOCATION GPRS and HLR in 3G) which sends *a profile with a QoS* in the response..

When the UE wants to connect to internet services (this is automatic in 4G as the UE are permanently connected), this is the QoS (*requested by mobile*) used in the Create Session Request 1)) sent to the PGW by the MME.

The PGW must ask the PCRF which QoS to allocate to the UE. The PCRF has a more detailed profile than in the HSS and the PCRF take actions following SDP business rules which can eventually modulate a user's profile following his actual usage and/or his account credit status. The PGW uses a request 2) on Figure 14.1, which is Credit Control Request (CCR) based on the DIAMETER protocol.

The PCRF returns the *Qos allocated by PCRF* in the Credit Control Ack 3). In a classical GGSN or PGW, this QoS is returned as such 5) to the UE (it is not recomputed by the PGW). *This QoS does not depend from the other users on the same cell.*

The 2 parameters *QoS requested by mobile* and *QoS allocated by PCRF* are passed by the GGSN to the PCEF. The PCEF is a « smart IP router » which follows the instructions to implement the QoS policy (it can act *on the Down Link bandwidth* by dropping packets from the SGi interface) transmitted by the PGW in 4). In case of packet losses, the TCP protocol in the mobile browsers will repeat the request, hence a slowdown of the user traffic The implementation of the Up Link bandwidth usage is the responsibility of the UE using the *QoS allocated by PCRF* without any change by the PGW.

In the Advanced PCC architecture of this chapter, the PGW has a Data Base BD URR of the Use of Radio resources. It contains for each cell 3G (UTRAN) or 4G (eUTRAN) the maximum bandwidth available based on the number of transmitters and technology. The PGW keeps in this Data Base the cell number and the *QoS allocated by the PGW* of each user.

Figure 14.1 shows how to both centralize the use of the bandwidth and have a large DPI processing capacity by using eventually several PCEF#1, PCEF#N controlled by a unique PGW handling all the GTP-C control messages used for creating, modifying or deleting sessions.

This centralization is necessary to handle the high 4G traffic (90 Gbps with 7 million Subscribers for a MNO in Taiwan which gives 2000 sessions / sec. Up to 10 PCEFs may be necessary; all controlled by a

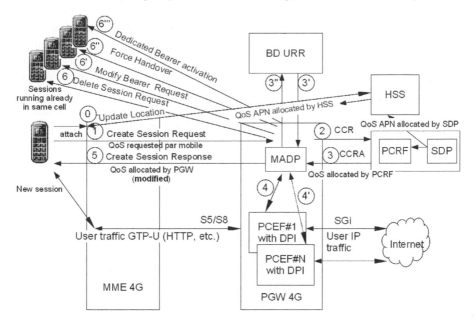

Figure 14.1 PGW 4G with Dynamic Pre-emptive Priorities.

single PGW.

For smaller configurations a single PCEF and the PGW may be integrated in the same server.

Figure 14.1 is the case of a new session while there are several users already in the cell (SAI). In 1) the UE has attached to the MME and the MME creates a session with the PGW by sending a Create Session Request (protocol GTP v2) to the PGW. In the 3G case3G it would be the SGSN which sends a Create PDP Context (protocol GTP v1) to the GGSN. More precisely the *Qos requested by mobile* (from the HSS) is sent in the « Bearer Context » parameter. The message is received by the PGW which passes 2) the demand in a CCR (Credit Control Request) message (Gx protocol) to the PCRF with the user identifications (IMSI) and the service requested (APN).

The PCRF has a rule data base (the SDP) which allows to return a *QoS allocated by PCRF* but *which does not account for the other users' occupation* of the same cell as the new entrant. The CCA message (Credit Control Ack) is returned by the PCRF 3). Up to now it is the classical PCC architecture.

In the advanced PCC architecture, the PGW has a MADP module, which interrogates the BD URR to know the occupation by *other users of* *the same cel*l. A knapsack problem is resolved. As explained above, it computes the *QoS allocated by PGW* which may require offloading other users. Possible but not shown on Figure 14.1, this new "QoS allocated by PGW" could as in the case of a mobile initiated Bearer Context modification, be sent to the PCRF using a CCR message.

The PCEFs are separate logical entities. The PGW knows the number of sessions and DL bandwidth C_k of the various PCEF#1, 2,..,N which it controls. It *selects a PCEF* which has a spare capacity, for example PCEF#1. The PGW sends 4) a GTP v2 message Create Session PCEF Request as below:

Create PCEF Session Request (PGW->PCEF)

```
_____ Header IP received _____
    Src IP '192.168.2.2' Src Port 2123
    Dst IP '192.168.0.4' Dst Port 2123
 - - Super Detailed GTP, Gi, SIP, DIAMETER, RADIUS Analyser RS=1 ENCAPS=1 msglen=213 - -
        (48)GTP version 2(1=GSM,2= LTE), T(presence TEID)=1
        Message Type:(244):Create PCEF Session Request (PGW->PCEF Vendor Specific
HALYS-Vedicis)
        Length_Payload 209
        TEID= B4B3B2B1(Hexa) 3031675569(Dec)
        SequenceNumber 12345
        Spare 00
          (1):IMSI(International Mobile Subscriber Identity)
            L(TLV type) = 8
            CR flag: 00 Instance: 00
            Data: 208103695303502
          (76):MSISDN
            L(TLV type) = 7
            CR flag: 00 Instance: 00
            Data: 33608123456
          (82):Radio Access Technology(RAT)
            L(TLV type) = 1
            CR flag: 00 Instance: 00
            Data: (1):UTRAN
          (86):User Location Information(ULI)
            L(TLV type) = 8
            CR flag: 00 Instance: 00
            SAI  present(7 otets)
             MCC 208 MNC 1 LAC 12345 SAC 1234
          (71):APN(Access Point Name)
            L(TLV type) = 12
            CR flag: 00 Instance: 00
            Data: xxx.yyy.zzz
          (99):PDN Type
            L(TLV type) = 1
            CR flag: 00 Instance: 00
```

PDN type: (1):IPv4
(79):PDN Address Allocation(IP allocated to UE)
 L(TLV type) = 5
 CR flag: 00 Instance: 00
 Data: 123.124.125.126
 PDN Type (1):IPv4
(127):APN Restriction
 L(TLV type) = 1
 CR flag: 00 Instance: 00
 Data: 12
(72):Aggregate Maximum Bit Rate(AMBR)
 L(TLV type) = 8
 CR flag: 00 Instance: 00
 Uplink: 70000 bits/sec
 Downlink: 200000 bits/sec
(78):Protocol Configuration Options(PCO)
 L(TLV type) = 23
 CR flag: 00 Instance: 00
(93):Bearer Context
 L(TLV type) = 31
 CR flag: 00 Instance: 00
 (73):EPS Bearer ID (EBI)
  L = 1
  CR flag: 00 Instance: 00
  Data: 5
 (80):Bearer level Quality of Service – /*allocated by the SGSN or MME from the HLR or*
  L = 22   *HSS QoS in the UPDATE LOCATION (GPRS)*/
  CR flag: 00 Instance: 00
  ARP
          Pre-emption Vulnerability: (0):Enabled
          Priority Level: 7
          Pre-emption Capability: (1):Disabled
  Label (QCI): (6):Non-GBR,priority = 7, 100 ms: Interactive gaming
  Maximum Bit Rate for Uplink: 1024 kbps
  Maximum Bit Rate for Downlink: 1024 kbps
  Guaranteed Bit Rate for Uplink:
                    0 kbps */* no minimum Bit Rate is requested by this
                    particular Internet service in the HLR or HSS QoS*/
  Guaranteed Bit Rate for Downlink: **0 kbps**
(87):Fully qualified End Point Identifier(F-TEID)
 L(TLV type) = 9
 CR flag: 00 Instance: 02
 IP type: (2):IPv4
 Interface type: (5):S5/S8 PGW GTP-U interface (set by PGW in Create Session
Response PGW->SGW)
 TEID/GRE Key: 00000002(Hexa) 2(Dec)
 IPv4: 192.168.0.4
(87):Fully qualified End Point Identifier(F-TEID)
 L(TLV type) = 9

CR flag: 00 Instance: 00
IP type: (2):IPv4
Interface type: (1):S1-U SGW GTP-U interface (added by SGW in Create Session Response SGW->MME)
TEID/GRE Key: 00000002(Hexa) 2(Dec)
IPv4: 192.168.0.4
(87):Fully qualified End Point Identifier(F-TEID)
L(TLV type) = 9
CR flag: 00 Instance: 00
IP type: (2):IPv4
Interface type: (0):S1-U eNodeB GTP-U interface (set by MME in Create Session Request MME->SGW)
TEID/GRE Key: A1A2A3A4(Hexa) 2711790500(Dec)
IPv4: 2.3.4.5
(87):Fully qualified End Point Identifier(F-TEID)
L(TLV type) = 9
CR flag: 00 Instance: 02
IP type: (2):IPv4
Interface type: (4):S5/S8 SGW GTP-U interface (added by SGW in Create Session Request SGW->PGW
TEID/GRE Key: A1A2A3A4(Hexa) 2711790500(Dec)
IPv4: 2.3.4.5

You notice that the guaranteed (minimum) Bit Rate is 0 in the QoS allocated from the HLR or HSS. This is the case for the less demanding internet services. For services such as VoIP the QoS profile would specify a sufficient QoS for the guaranteed Bit Rate. If the value is 0 it is left to the PCRF to allocate the resources.

Create PCEF Session Response (PCEF->PGW)

- - Super Detailed GTP, Gi, SIP, DIAMETER, RADIUS Analyser RS=0 ENCAPS=1 msglen=53 -
-
(48)GTP version 2(1=GSM,2= LTE), T(presence TEID)=1
Message Type:(245):Create PCEF Session Response(PGW<-PCEF Vendor Specific HALYS-Vedicis)
Length_Payload 49
TEID= B4B3B2B1(Hexa) 3031675569(Dec)
SequenceNumber 12345
Spare 00
(2):Cause
L(TLV type) = 2
CR flag: 00 Instance: 00
Data: (16):Request Accepted
(93):Bearer Context
L(TLV type) = 31
CR flag: 00 Instance: 00
(73):EPS Bearer ID (EBI)
L = 1

CR flag: 00 Instance: 00
Data: 5
(80):Bearer level Quality of Service - /*raw QoS allocated by the PCRF without considering
L = 22   *the other users in the ULI MCC 208 MNC 1 LAC 12345 SAC 1234*/*
CR flag: 00 Instance: 00
ARP
Pre-emption Vulnerability: (0):Enabled
Priority Level: 7
Pre-emption Capability: (1):Disabled
Label (QCI): (6):Non-GBR,priority = 7, 100 ms: Interactive gaming
Maximum Bit Rate for Uplink: 1024 kbps
Maximum Bit Rate for Downlink: 1024 kbps
Guaranteed Bit Rate for Uplink: 256 kbps
Guaranteed Bit Rate for Downlink: 768 kbps

For the Downlink (DL), the traffic control is *performed by the PCEF* which throttles the SGi interface:

- the *QoS allocated by PGW* (may be different from *QoS allocated by PCRF*) based on congestion of the SGi interface.
- the GTP-U Tunnel Identifiers which allow the *selected PCEF* to route the DL user traffic to the eNodeB (to the SGSN for 3G)

For the Uplink (UL) the traffic control is *performed by the UE*. The selected PCEF returns 4) a Tunnel Identifier so that the eNodeB may send the UL traffic to this selected PCEF#1.

A Create Session Response is sent 5) to the MME. As a result the user traffic of the new entrant *starts only when the selected PCEF has been readied by 4)*. Immediately after sending 5) the PGW may send Delete Session Requests 6), Forced Handover 6") to offload a non priority user to a neighbor cell. As the BD URR with latitude-longitude of the center of cells, it allows to find the cells which are close to the one for which capacity is needed. Also it could (but this is against the policy « if service granted,   full   subscription   capacity »)   send   Modify   Bearer Request commands 6') to reduce the QoS of a user, or change the traffic to secondary PDP Context 6''') as detailed in section 14.3.5.

14.3.4 Including the Maximum Processing Bandwidth of the SGi (or Gi in 3G) of the PCEFs

It is possible to include also the internet bandwidth limit (typically 30 Gbits / sec for a large of PGW or GGSN). We have 2 capacity constrains:

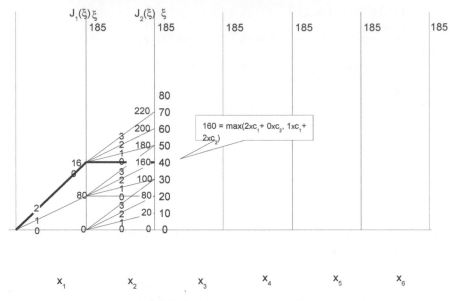

Figure 14.2 Bellman-Jacobi Dynamic Programming iteration for the knapsack problem.

The Bellman-Jacobi type dynamic programming algorithm uses N state vectors. The dimension M= MIN / min a_i, i.e. M= 185/1 = 185 in the example illustrated by Figure 14.2. One sees that the intermediate state $\zeta =$ 40 at step 2 is reached with an optimal trajectory $x_1 = 2$, $x_2 = 0$ with an optimum $J_2(40) = 160$. State $\zeta = 40$ is reached with 2 trajectories: $x_1 = 2$, $x_2 = 0$ or $x_1 = 1$, $x_2 = 2$ and the optimum is reached with the first trajectory.

This is a very fast algorithm as it is linear in the number of steps (the number of different QoS policies).

14.3.5 Use of Secondary PDP Context for the Offloading of Non-Priority Users

14.3.5.1 Purpose of the Secondary PDP Contexts

Secondary PDP Contexts were introduced to handle such services as VoIP. In the idle phase with QoS defined by a Primary PDP Context, a small bandwidth is sufficient to wait for the SIP message for an incoming call or a call generated by the SIP client of the UE. Depending on the service it is necessary to have a better QoS with the same IP address of the terminal:

Medium-low	QoS if it is a voice call	the 2nd PDP Context,
Medium-high	QoS if it is a ciphered voice call	the 3rd PDP Context
High	QoS if it is a visio call	the 4th PDP Context.

This way no bandwidth is wasted in the idle phase or if it is a normal voice call.

Secondary PDP Contexts have the same APN as the main PDP Context and there is no reallocation of the UE IP address when a secondary PDP context is activated. If such service is provisioned in the HSS (or HLR in 3G), the HSS sends an INSERT_SUBSRIBER_DATA which contains a GPRS subscription Data parameter with a list of PDP Contexts sharing the same IPN and different QoS. Their order in the list defines the main PDP Context the 2nd, the 3rd, the 4th, etc.

The change of PDP Context is called a Secondary PDP Context activation in 3G and "Dedicated Bearer Activation" in 4G. It can be mobile initiated (the above VoIP example) of network initiated by the PCRF.

In general the Primary PDP Context is to be used for an idle and has a lower priority than the secondary PDP Contexts. In this chapter, we use it the other way. For low value subscribers, the Primary PDP Context of the Internet service will have an average QoS, but a secondary PDP Context of the Internet service has a lower QoS. This allows the pre-emptive priority management to be more flexible:

a) delete the low priority subscriber,
b) force a HANDOVER to a neighboring cell provided the MME supports it,
c) reduce his QoS by switching him to a secondary PDP context as explained in this section.

The Figure 14.3 explains the network initiated change of Qos for the pre-emptive priority. It could be decided by the PCRF or by the PGW who has the knowledge of the cell bandwidth usage.

The PGW will send a Create Bearer Request 1) which refers (same correlation ID) to the main PDP context (this has a sense only if the session was already created with the allocation on an IP address to the UE. If the change is initiated by the PCRF it sends an IP-CAN Session Modification message on the Gx interface.

This procedure has an exact equivalent in 3G, the GPTv1 Initiate Context Activation Request is used instead of a Create Bearer Request.

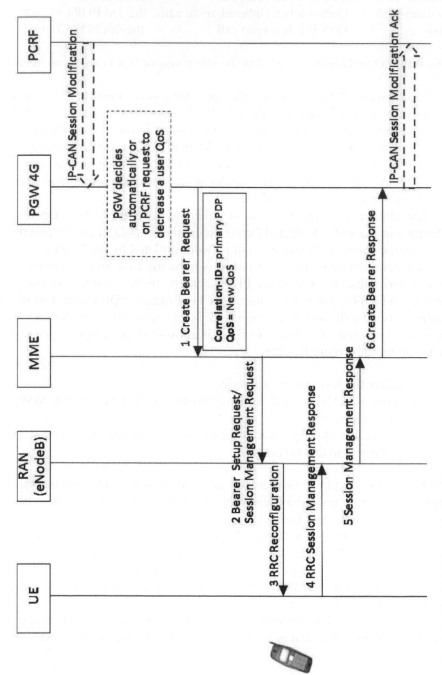

Figure 14.3 Dedicated Bearer Activation Procedure.

14.3.5.2 Extension of the Mathematical Model to Handle the Secondary PDP Case

For each class of subscribers we define the number of policies with the primary QoS x_{i1} and with secondary QoS x_{i2}. The monetization values are c_{i1} and c_{i2}, the bandwidths are a_i and a_{i2}. The new optimisation problem becomes:

$$\max P = \sum_i c_{i1} x_{i1} + c_{i2} x_{i2}$$

$$\sum_i a_{i1} x_{i1} + a_{i2} x_{i2} <= \text{MIN}, \text{ single state variable } \zeta.$$

$$x_{i1} + x_{i2} <= d_i, \ x_{i1} \text{ and } x_{i2} \text{ integer}$$
$$0 <= x_{i1}, x_{i2}$$

This is not quite a "basic" knapsack problem but as it has only one state variable ζ, the dynamic programming algorithm of section 14.3.4 applies.

14.3.5.3 More Complicated Case: Secondary PDP and Forced HANDOVER Capabilities

With forced HANDOVER the class # i service may be provided on the nominal cell j *but also from the neighbor cells* j_k, with k = 1..N, and also the *change to the secondary PDP is possible*. The index k= 0 means the nominal cell index.

Let x_{i1k} and x_{i2k} be the variables giving the number of class # i services provided with PDP 1 and 2 on the nominal cell or the neighbor cell k.

Let be R_k the *spare capacity* on the neighbor cell j_k . To

Let $\text{MIN}_k = \min (Bj_k, C_s - \text{other traffic of PCEF# s})$ (in general $\text{MIN}_k = Bj_k$ as in 14.3.4)

To limit the optimization problem size assume that the PCC may force a HANDOVER only in the nominal cell j, and we do not change the QoS of the users in the neighbor cell. The optimization problem becomes:

$$\max P = \sum_i \sum_k c_{i1} x_{i1k} + c_{i2} x_{i2k} - K \sum_i \sum_{k \neq 0} x_{i1k} + x_{i2k} , \ K \text{ some} >0 \text{ constant}$$

$$\sum_i a_{i1} x_{i10} + a_{i2} x_{i20} \qquad <= \text{MIN, state variable } \zeta , \text{ nominal cell}$$

$$\sum \sum a_{i1} x_{i1k} + a_{i2} x_{i2k} <= \text{MIN}_k, \text{ state variable } \zeta_k, k = 1..N , \text{ neighbor cells}$$
$$i \ k \neq 0$$

$$\sum x_{i1k} + x_{i2k} <= d_i, \text{ means that the demand on policy # i can be satisfied on}$$

the neighbor cells or partially with a secondary PDP context
$0 \leq x_{i1k}, x_{i2k}, x_{i1k}$ and x_{i2k} integer

The term $-K \Sigma\Sigma \ x_{i1k} + x_{i2k}$ is a *penalization function*. It penalizes the value of the optimum if there any handover on the neighbor cells ($k \neq 0$) is performed. *The optimal solution will then have a minimum number of handovers.*

To his problem has now several inequality constraints, but the number of neighbor cells will be less than 10 and the number of variables is then less than 100. A Bellman multi-state variable dynamic programming algorithm is then quite fast in this larger case also.

14.3.5.4 Accounting for the Power Measurements of the Neighboring eNodeB

An eSMLC (Chapter 13) with the LPPa (LTE Positioning Protocol A) could be useful for a definition of a neighbor cell better than a static definition from the geographic location of the center of the cells. Besides providing an accurate location of the new entrant, power measurements, see examples in [0.8 chapter 6], of the various neighboring eNodeBs may be provided by an eSMLC and only neighbor cells j_k with sufficient power measured by the mobile will be considered. To model this, instead of a uniform penalization K for a handover to a neighbor cell k, the measured received power P_{tk} each mobile t candidate to be handed over to a neighbor cell k is used. The model is more complicated with many more variables (one for each subscriber of the nominal cell, not one per policy) and left to the reader as we must use a new penalization which tends to avoid a handover to a neighbor cell if the received power is low.

To end this section, one sees that in order to be able to perform this dynamic preemptive priority optimization, the PCC must be aware of all cell changes as in 13.4.2.3. That is the "location information change" must be activated: this is available if the MMEs have at least Release 10 of [3.8] for at least the simplest policies, release which starts "Advanced LTE". To have the possibilities of forced handovers, one has to wait for future releases of proprietary implementations.

14.4 Roles of the PCRF and of the SDP

In order to have a full understanding of the preemptive priority system, what is the real usefulness of a PCRF? In Figure 14.1:

QoS requested by mobile = QoS APN allocated by HSS = QoS APN allocated by SDP

This common QoS defines both the Uplink and Downlink traffic and is known both of the PCRF and of the MME. *If the HSS is individually updated for each subscriber's profile change*, by the SDP, not only for a generic APN, the PCRF is not necessary for just the pre-emptive priority. Whether the SDP is embedded in an MNO's customer care system, or separate (case of a MNVE or full MVNO), the HSS is slave to the HSS to receive the profiles. The PCRF role remains nevertheless to process the traffic volume events sent by the PGW for charging purposes, and also to send to the PGW dynamic QoS change requests coming from external IMS Application Servers such as VoIP, or from external web applications(automatic credit reload).

When we write Qos requested by mobile = QoS APN allocated by HSS, we mean that the value of the priority and bandwidth parameters is the same but their coding is different, the MME performing a conversion.

Then if one needs only to control the QoS with preemptive priorities without a charging function, the SDP is sufficient to initialize the HSS without a PCRF. In this simplified case, the PGW does not interrogate the PCRF and uses only the « QoS requested by mobile » which is QoS-subscribed of the HSS in order to perform the knapsack problem.

The PCRF adds to the cost of the PCC but adds a dynamic charging and service control function, which will become a strategic tool for bandwidth monetization in the context of emerging "Over The Top" applications.

References and Further Reading

[14.1] ETSI TS 123 203 V11.11.0 (2013-09), "Universal Mobile Telecommunications System (UMTS); LTE; Policy and charging control architecture, Release 11"

[14.2] R. Bellmann, « Dynamic Programming », Princeton University Press, 1957

[14.3] A.Henry-Labordère, S.Cruaux, "Passerelle d'accès des mobiles à internet avec gestion préemptive des priorités", Patent FR 2013 ("Access Gateway of mobiles to internet services with pre-emptive priority management").

[14.4] ETSI TR 136 902 V9.3.1 (2011-05), "LTE;Evolved Universal Terrestrial Radio Access Network (E-UTRAN);Self-configuring and self-optimizing network (SON) use cases and solutions (3GPP TR 36.902 version 9.3.1 Release 9) ".

[14.5] ETSI TS 124 008 V11.8.0 (2013-10), "Digital cellular telecommunications system (Phase 2+);Universal Mobile Telecommunications System

(UMTS);LTE;Mobile radio interface Layer 3 specification; Core network protocols; Stage 3 (3GPP TS 24.008 version 11.8.0 Release 11)".

[14.6] ETSI TS 123 401 V11.7.0 (2013-09) "LTE; General Packet Radio Service (GPRS) enhancements for Evolved Universal Terrestrial Radio Access Network (E-UTRAN) access (3GPP TS 23.401 version 11.7.0 Release 11)".

Abbreviations and Acronyms

A3	Authentication algorithm A3 used in 2G GSM with a challenge between the HLR and the SIM card using a random number and a result computed with a shared secret Ki between the HLR (AuC) and the SIM card.
A38	A single algorithm performing the functions of A3 and A8
A5/1	Encryption algorithm A5/1
A5/2	Encryption algorithm A5/2
A5/X	Encryption algorithm A5/0-7
A8	Ciphering key generating algorithm A8
AA19	Standard GSM contract between two operators for the charging of the SMS-MT sent to their own subscribers by the other.
AB	Access Burst
AC	Access Class (C0 to C15)
	Application Context
ACC	Automatic Congestion Control
ACCH	Associated Control CHannel
ACK	ACKnowledgement
ACM	Accumulated Call Meter (a zone of a SIM card)
	Address Complete Message (Response to a ISUP Call setup)
ACMmax	Maximum of the Accumulated Call Meter
ACSE	Association Control Service Element
ACU	Antenna Combining Unit
ADC	ADministration Center
	Analogue to Digital Converter
ADD	Automatic Device Detection (inclusioj of IMEIsv in Update Location)
ADN	Abbreviated Dialing Number
ADPCM	Adaptive Differential Pulse Code Modulation
AE	Application Entity
AEC	Acoustic Echo Control
AEF	Additional Elementary Functions
AGCH	Access Grant Channel

AKA	Authentication and Key Agreement
Ai	Action indicator
AMPS	Advanced Mobile Phone System (Analog mobile Radio system)
ANSI	American National Standards Institute
AoC	Advice of Charge
AoCC	Advice of Charge Charging supplementary service
AoCI	Advice of Charge Information supplementary service
APLMN	Associated Public Land Mobile Network
APN	Access Provider Name
ARP	Allocation and Retention Priority (Allows to setpreemptive priorities for the Creation of data sessions) Alternative Roaming Providers for the data roaming (Notion introduced in the new European regulation for the Local Break-Out)
ASE	Application Service Element
AS	Autonomous System: a subset of the IP network with a common routing policy.
ASN	Autonomous System Number, a unique 16 bits number(IPV4) or 32 bits number(IPV6) defining the set of IP addresses of an operator for the purpose of configuring his "Border Gateway" and included in his IR21 document. The ASN is delivered at the same time as some public IP addresses are allocated to companies.
ASN.1	Abstract Syntax Notation One
ARFCN	Absolute Radio Frequency Channel Number
ARQ	Automatic ReQuest for retransmission
ASP	Application Service Provider (Content Provider for Internet services)
ATT (flag)	ATTach
AU	Access Unit
AuC	Authentication Center
AUT(H)	AUThentication
AVP	Attribute Value Pair, name of the parameters in the RADIUS and DIAMETER protocols
BA	BCCH Allocation
BAIC	Barring of All Incoming Calls supplementary service
BAOC	Barring of All Outgoing Calls supplementary service
BBERF	Bearer Binding and Event Reporting Function, function in the SGW responsible for bearer binding.
BCC	Base Transceiver Station (BTS) Colour Code

BCCH	Broadcast Control CHannel
BCD	Binary Coded Decimal
BCF	Base station Control Function
BCIE	Bearer Capability Information Element
BER	Bit Error Rate
BEREC	Body of European Registrators for Electronic Communications
BFI	Bad Frame Indication
BGP	Border Gateway Protocol
BGW	Border Gateway. Connects a PLMN to GRX
BI	all Barring of Incoming call supplementary services
BIB	Backward Indicator Bit
BIC-Roam	Barring of Incoming Calls when Roaming outside the home PLMN country supplementary service
Bm	Full-rate traffic channel
BN	Bit Number
BO	all Barring of Outgoing call supplementary services
BOIC	Barring of Outgoing International Calls supplementary service
BOIC-exHC	Barring of Outgoing International Calls except those directed to the Home PLMN Country supplementary service
BS	Basic Service (group)
	Bearer Service
BSG	Basic Service Group
BSC	Base Station Controller GSM 2G
BSIC	Base transceiver Station Identity Code
BSIC NCELL	BSIC of an adjacent cell
BSN	Backward Sequence Number
BSS	Base Station System (GSM 2G)
BSSAP	Base Station System Application Part
BSSAP-LE	BSSAP with Location Extension
BSSMAP	Base Station System Management Application Part
BSSOMAP	Base Station System Operation and Maintenance Application Part
BTS	Base Transceiver Station GSM 2G
C	Conditional
C-PDS	CDMA Packet Data Service
CA	Cell Allocation
CAI	Charge Advice Information
CAMEL	Customised Application for Mobile Network

CAMELize	Action to provision a CAMEL record in a subscriber's profile for other purpose than pre-payment in order to trigger an IN service, such as automatic APN correction.
CB	Cell Broadcast
CBC	Cell Broadcast Center
	Ciphering Block Chaining (used in the 3DES algorithm)
CBCH	Cell Broadcast CHannel
CBMI	Cell Broadcast Message Identifier
CC	Country Code
	Call Control
CCA	Credit Control Acknowledgement (Diameter Credit Control Application)
CCBS	Completion of Calls to Busy Subscriber supplementary service
CCCH	Common Control CHannel
CCF	Conditional Call Forwarding
CCH	Control CHannel
CCITT	Comité Consultatif International Télégraphique et Téléphonique (The International Telegraph and Telephone Consultative Committee)
CCM	Current Call Meter
CCP	Capability/Configuration Parameter
CCPE	Control Channel Protocol Entity
CCR	Credit Control Request (Diameter Credit Control Application)
Cct	Circuit
CDMA	Code Division Multiple Access
CDR	Call Detailed Record (billing record)
CDUR	Chargeable DURation
CED	called station identifier
CEIR	Central Equipment Identity Register
CEND	end of charge point
CEPT	Conférence des administrations Européennes des Postes et Telecommunications
CF	Conversion Facility
	all Call Forwarding services
CFB	Call Forwarding on mobile subscriber Busy supplementary service
CFNRc	Call Forwarding on mobile subscriber Not Reachable supplementary service
CFNRy	Call Forwarding on No Reply supplementary service

CFU	Call Forwarding Unconditional supplementary service
CHP	CHarging Point
CHV	Card Holder Verification information
CI	Cell Identity
CIC	Circuit Identification Code
CIR	Carrier to Interference Ratio
CIS	Critically Important Services(administrations, utilities)
CKSN	Ciphering Key Sequence Number
CLI	Calling Line Identity
CLIP	Calling Line Identification Presentation supplementary service
CLIR	Calling Line Identification Restriction supplementary service
CM	Connection Management
CMD	CoMmanD
CMM	Channel Mode Modify
CNF	CoNFirmation (Answer to a REQ(Request))
CNG	CalliNG tone
CNTR	Counter (control for OTA SIM security)
COLI	COnnected Line Identity
COLP	COnnected Line identification Presentation supplementary service
COLR	COnnected Line identification Restriction supplementary service
COM	COMplete
COMP 128	Authentication and Ciphering algorithm used for A3 and A8 (GSM)
COMP 128-2	Improved algorithm used for UMTS
CONNACK	CONNect ACKnowledgement
C/R	Command/Response field bit
CRC	Cyclic Redundancy Check (3 bit)
CRE	Call RE establishment procedure
CRX	CDMA Roaming ExChange (equivalent of GRX for CDMA data roaming)
CS	Domain Circuit Service Domain (includes MSCs)
CSPDN	Circuit Switched Public Data Network
CT	Call Transfer supplementary service
	Channel Tester
	Channel Type
CTR	Common Technical Regulation
CUG	Closed User Group supplementary service

CW   Call Waiting supplementary service
DA   Destination Address
DAC   Digital to Analogue Converter
DAMPS   Digital AMPS (the TDMA mobile adio system)
DB   Dummy Burst
DCCH   Dedicated Control CHannel
DCE   Data Circuit terminating Equipment
DCF   Data Communication Function
DCN   Data Communication Network
DCS1800   Digital Cellular System at 1800MHz
DES   Data Encryption Standard(used for security of OTA SIM)
3DES   Triple DES (with 2 keys)
DET   DETach
DHCP   Dynamic Host Configuration Protocol . The DHCP server
                assigns an IP address to a terminal through this protocol.
DIAMETER   an Authenticatio, Authorisation and Charging Protocol
                which is more comprehensive than RADIUS
DISC   DISConnect
DL   Data Link (layer)
DLCI   Data Link Connection Identifier
DLD   Data Link Discriminator
Dm   Control channel (ISDN terminology applied to mobile
                service)
DMR   Digital Mobile Radio
DNIC   Data network identifier
DP   Dial/Dialed Pulse
                Destination Point (of an IN service)
DPC   Destination Point Code
DRX   Discontinuous reception (mechanism)
DSE   Data Switching Exchange
DSI   Digital Speech Interpolation
DSS1   Digital Subscriber Signaling No1
DTAP   Direct Transfer Application Part
DTE   Data Terminal Equipment
DTMF   Dual Tone Multi Frequency (signaling)
DTX   Discontinuous transmission (mechanism)
DVB-H   Digital Video Broadcast-Handheld (MULTICAST method
                for mobile TV)
EA   External Alarms
EAL   Evaluation Assurance Level

EAL1	A low but cheap level of security, system functionally tested
EAL5	A high level of certification for secured communication systems semiformally designed and tested.
EAL7	Highest level for government, formally verified designed and tested
EAP	Extensible Authentication Protocol
EAP-AKA	EAP for 3rd Generation USIM cards
EAP-SIM	EAP for SIM cards
EBSG	Elementary Basic Service Group
ECM	Error Correction Mode (facsimile)
Ec/No	Ratio of energy per modulating bit to the noise spectral density
ECT	Explicit Call Transfer supplementary service
EDGE	Enhanced Data Rates for GSM Evolution ("2.75 intermediate generation allowing much higher rates without UMTS: 170 kbits/sec for visio)
EEL	Electric Echo Loss
EIA	Electronic Industries Equipment
EIR	Equipment Identity Register
EL	Echo Loss
EMC	ElectroMagnetic Compatibility
eMLPP	enhanced Multi-Level Precedence and Pre-emption service
EMMI	Electrical Man Machine Interface
ENUM	E164 Number Mapping
ePDG	evolved Packet Data Gateway, Gateway with an AAA server to secure a non LTE access (e.g. WiFi) to the LTE PGW with the S2b interface.
EPROM	Erasable Programmable Read Only Memory
ERP	Ear Reference Point
	Equivalent Radiated Power
ERR	ERRor
ESME	External Short Message Entity (an ASP or ISP) connected by SMPP
ESN	Electronic Serial Number
ESP	Encapsulating Security Payload
ETR	ETSI Technical Report
ETS	European Telecommunication Standard
ETSI	European Telecommunications Standards Institute

E164	Format of the "ordinary" telephone numbers with a "Country Code"(CC) and a Network Destination Code(NDC)
E212	Format of the "IMSI" telephone numbers with a "Mobile Country Code"(MCC) and a Mobile Network Code(MNC)
E214	Format of a Destination Address, a mix of E164 and E212
FA	Full Allocation
Fax	Adaptor
FAC	Final Assembly Code
FACCH	Fast Associated Control CHannel
FACCH/F	Fast Associated Control Channel/Full rate
FACCH/H	Fast Associated Control Channel/Half rate
FB	Frequency correction Burst
FCCH	Frequency Correction CHannel
FCS	Frame Check Sequence
FDM	Frequency Division Multiplex
FDN	Fixed Dialing Number
FEC	Forward Error Correction
FER	Frame Erasure Ratio
FH	Frequency Hopping
FIB	Forward Indicator Bit
FISU	Fill In Signal Units
FN	Frame Number
FR	Full Rate
FSG	Foreign Subscriber Gateway
FSN	Forward Sequence Number
ftn	forwarded-to number
GBR	Guaranteed Bit Rate
GCR	Group Call Register
GGSN	GPRS Gateway Support Node, in GPRS equipped network, provides the interface between an operator's own IP network and the external IP network (GRX mostly)
Gi	Interface to Internet from/to GGSN with normal IP packets, not encapsulated
GLR	Gateway Location Register. Acts as a HLR for the visitor to avoid UPDATE LOCATION being sent to the HPLMN. Used by a VPLMN to maintain a visitor in its own network.
GMLC	Gateway Mobile Location Center
GMSC	Gateway Mobile-services Switching Center
GMSK	Gaussian Minimum Shift Keying (modulation)
GPA	GSM PLMN Area

GPRS	General Packet Radio Service
GRX	The Intranet IP network used by mobile operators to exchange GPRS data. It is operated on a cooperative basis by the main international carriers.
GSA	GSM System Area
GSM	Global System for Mobile communications
GSM MS	GSM Mobile Station
GSM PLMN	GSM Public Land Mobile Network
GSM-R	GSM Railway adaptation (fast mobility)
GT	Global Title (E164 numbering address)
GTT	Global Title Translation
GUI	Graphic User Interface
Gx	Interface (protocol DIAMETER Credit Control) between PCEF and PCRF
Gy	Real-time charging by the GGSN using a DIAMETER interface with the SDP
Gz	Off-line charging interface of the GGSN which transfers data tickets with the GTP' protocol
H-PCRF	Implementation of the PCRF in the HLPMN (Home access)
H-GMLC	Home Gateway Mobile Location Center IP address, used in 4G Location services.
H223	Multplexing protocol for visio, voice and control
H245	Control protocol in H223 for H324-M communications
H263, H264	visio encoding standard for H324-M
H324-M	Standard for visio calls 3G using 64 kbits and ISUP
HANDO	HANDOver
HDLC	High level Data Link Control
HLC	High Layer Compatibility
HLR	Home Location Register
HMAC	Hash Message Authentication
HMAC-MD5	A signature algorithm returning 16 octets for a string of any length. It uses a shared secret between the two entities exchanging messages
HOLD	Call hold supplementary service
HPLMN	Home PLMN
HPU	Hand Portable Unit
HR	Half Rate
HSN	Hopping Sequence Number
HSPDA	High Speed Downlink Packet Access (gives 250 Kbits useful data for visio)

HU	Home Units
I	Information frames (RLP)
IA	Incoming Access (closed user group SS)
IA5	International Alphabet 5
IAM	Initial Address Message
IAP	Internet Access Provider (Provides the access a modem or a permanent IP connection to the Internet, not necessarily a Portal or Content Provider)
IC	Interlock Code (CUG SS)
ICB	Incoming Calls Barred (within the CUG)
ICC	Integrated Circuit(s) Card
IC(pref)	Interlock Code of the preferential CUG
ICM	In-Call Modification
ID	IDentification/IDentity/IDentifier
IDN	Integrated Digital Network
IE (signaling)	Information Element
IEC	International Electrotechnical Commission
IEI	Information Element Identifier
IETF	Internet Engineering Task Force
I-ETS	Interim European Telecommunications Standard
IGP	International Gateway Provider (SCCP access to the SS7 network)
IGW	International SCCP Gateway (synonym of IGP).
IMEI	International Mobile station Equipment Identity
IMEISV	International Mobile station Equipment Identity with software version
IMS	IP Multimedia System
IMSI	International Mobile Subscriber Identity
IN	Interrogating Node
IN	Intelligent Network
IN service	Service such as pre-payment, number correction, APN correction performed by a SCP
INAP	Intelligent Network Application Part
InitialDP CAMEL	service to start an IN Service
IOT	Inter Operator Tariff
IP	Internet Protocol
IP-PDP	Address IP allocated by the GGSN in the response to a Create PDP Context
IP-TE-VPN	Address IP subsequently allocated by a RADIUS server to create a secure VPN tunnel

IR21	International Roaming 21document(description of the detailed numbering plan as standardized by the GSM association)
ISC	International Switching Center
ISD	Abbreviation for INSERT SUBSRIBER DATA
ISDN	Integrated Services Digital Network
ISO	International Organization for Standardization
ISP	Internet Service Provider (Content Provider for Internet services)
ISUP	ISDN User Part (of signaling system No.7)
ITC	Information Transfer Capability
ITU	International Telecommunication Union
IVR	Interactive Voice Response
IWF	InterWorking Function (used for Circuit Mode internet access, Modems V110 -> packet converter)
IWMSC	InterWorking MSC
IWU	InterWorking Unit
k	Windows size
K	Constraint length of the convolutional code
Kc	Ciphering key
Ki	Individual subscriber authentication key
KiC	Key for Ciphering (OTA SIM)
KiD	Key for RC/CC/DS signature (OTA SIM)
KiK	Key for protecting KiC and KiD
L1	Layer 1
L2ML	Layer 2 Management Link
L2R	Layer 2 Relay
L2R BOP	L2R Bit Orientated Protocol
L2R COP	L2R Character Orientated Protocol
L3	Layer 3
LA	Location Area
LAC	Location Area Code
LAI	Location Area Identity
LAN	Local Area Network
LAPB	Link Access Protocol Balanced
LAPDm	Link Access Protocol on the Dm channel
LBO	Local Break-Out, allows local subscribers of an ARP (VPLMN) to use the data services.
LBS	Location Based Services
LCN	Local Communication Network
LCS-AP	LCS Application Protocol

LCS	Location Services
LCSC	LCS Client
LCSS	LCS Server
LE	Local Exchange
LEMF	Law Enforcement Monitoring Facility
LI	Length Indicator
	Line Identity
	Lawful Interception
LLC	Low Layer Compatibility
Lm	Traffic channel with capacity lower than a Bm
LMSI	Local Mobile Station Identity
LMU A	Location Measurement Unit with only Air interface
LMU B	Location Measurement Unit integrated in a BTS.
LND	Last Number Dialed
LNP	Local Number Portability[2]
LPLMN	Local PLMN
LPP	LTE Positioning Protocol
LPPa	LTE Positioning Protocol Annex
LR	Location Register
LSSU	Link Status Signal Units
LSTR	Listener SideTone Rating
LTE	Local Terminal Emulator
LTE	Long Term Evolution (the new radio access standard of the 4G)
LU	Local Units
	Location Update
LV	Length and Value
M	Mandatory
MA	Mobile Allocation
MAC address	Media Access Control address. Unique identifier (manufacturer dependent) assigne to a physical network interface.
MACN	Mobile Allocation Channel Number
MAF	Mobile Additional Function
MAH	Mobile Access Hunting supplementary service
MAI	Mobile Allocation Index
MAIO	Mobile Allocation Index Offset
MAP	Mobile Application Part, there is MAP GSM and MAP IS-41(CDMA)
MBMS	Multimedia Broadcast Multicast Services (TV broadcast for expel on a single frequency per cell)

MC	Message Center (in the IS 41 network, equivalent of a GSM SMSC)
	Multi-Call (simultaneous bearer services)
MCC	Mobile Country Code
MCI	Malicious Call Identification supplementary service
MD	Mediation Device
MD5	A signature algorithm returning 16 octets for a string of any length including a shared secret at the end
MDL (mobile)	Management (entity) Data Link (layer)
MDN	Mobile Destination Number (IS 41)
ME	Maintenance Entity
	Mobile Equipment
MEF	Maintenance Entity Function
MF	MultiFrame
	Mediation Function
MFSMWR	MTU_FORWARD_SHORT_MSG_WWW_REQ
MFSMC	MTU_FORWARD_SHORT_MSG_CNF
MFSMHI	MTU_FORWARD_SHORT_MSG_HO_IND
MFSMWRSP	MTU_FORWARD_SHORT_MSG_WWW_RSP
MGT	Mobile Global Title
MHS	Message Handling System
MIB	Master Information Block, standard description of a system to be supervised by the SNMP protocol
MIC	Mobile Interface Controller
Milenage	A more secure equivalent 3G of the A5/A3 algorithm
MIP	Mobile IP
MLC	Mobile Location Center
MLU	Mobile Location Units
MM	Man Machine
Mobility	Management
MMD	Multi Media Domain
MME	Mobile Management Entity
MMI	Man Machine Interface
MMS	Multimedia Messaging Service
MM1	Protocols using IP standards(http) to exchange MMS between the cellphone and the MMSC
MM4	In the MMS architecture, protocol to send MMS from one MMSC to an other (interconnection), basically SMTP (e-mail).
MM5	In the MMS architecture, protocol to interrogate the HLRs

MM7	Protocols using IP standards so Content Provider sends MMS to an MMSC
MMT	Mobile Money Transer
MNC	Mobile Network Code
MNO	Mobile Network Operators
MNP	Mobile Number Portability
MO	Mobile Originated
MOC	Mobile Originated Call
MO-LR	Mobile Originating Location Request
MoU	Memorandum of Understanding
MPC	Mobile Positioning Center
MPEG 4	visio encoding standard for H324-M (alternative to H263 or H264)
MPH (mobile)	Management (entity) PHysical (layer) [primitive]
MPTY	MultiParTY (Multi ParTY) supplementary service
MRP	Mouth Reference Point
MS	Mobile Station
MSC	Mobile services Switching Center, Mobile Switching Center
Anchor MSC	Mobile Switching Center that is the first to assign a traffic channel to an MS
Serving MSC	MSC wich currently has the MS obtaining service at one of its cell sites
Tandem MSC	Previous Serving MSC in the handoff chain
MSCM	Mobile Station Class Mark
MSCU	Mobile Station Control Unit
MSISDN	Mobile Station International ISDN Number
MSRN	Mobile Station Roaming Number
MSRP	Message Session Relay Protocol, used with RCS to relay realetd instant messages and for file transfer
MSU	Message Signal Units
MT	Mobile Terminated
MT (0,1,2)	Mobile Termination
MT-LR	Mobile Terminating Location Request
MTC	Mobile Terminated Call
MTM	Mobile-To-Mobile (call)
MTN	Maintenance Regular Message
MTP	Message Transfer Part
MTP2	MTP Layer 2 (Link control level)
MTP3	MTP Layer 3 (Nework control sub-level (handles Point Codes)

MU	Mark Up
MULTICAST	"MULtiple broadCAST", broadcast such as what is used in Numerical TV (DVB-H) . The content is broadcasted to any number of receivers like in classical TV thus much more economical than UNICAST currently used.
MUMS	Multi User Mobile Station
MVNO	Mobile Virtual Network Operator
MWD	Message Waiting Data (indication in an HLR)
M2PA	MTP2 Peer-to-Peer Adaptation Layer
M2UA	MTP2 User Adaptation Layer
M3UA	MTP3 User Adaptation Layer
N/W	Network
NAMPS	Narrow AMPS
NAPTR	Name Authority Pointer: a record in a DNS
NAS	Network Access System: a Hot Spot, a SGSN or previously a modem rack for the data service using circuits.
NAT	Network Address Translation (a single external address for many local addresses)
NB	Normal Burst
NBIN	A parameter in the hopping sequence
NCC	Network (PLMN) Country Code
NCELL	Neighboring (of current serving) Cell
NCH	Notification CHannel
NDC	Network Destination Code
NDUB	Network Determined User Busy
NE	Network Element
NEF	Network Element Function
NET	Norme Europeenne de Télécommunications
NF	Network Function
NFC	Near Field Technology (contactless communication of handset)
NGN	Next GeNeration (IP based equipments and networks)
NI	Network Indicator
NIC	Network Independent Clocking
NI-LR	Network Induced Location Request
NM	Network Management
NMC	Network Management Center
NMSI	National Mobile Station Identification number
Node B	The UMTS 3G equivalent of a BTS (GSM 2G)
NPI	Number Plan Indentifier
NPS	Network Planning System

NSAP	Network Service Access Point
NSS	Network Vendors
NT	Network Termination
	Non Transparent
NTAAB	New Type Approval Advisory Board
NUA	Network User Access
NUI	Network User Identification
NUP	National User Part (SS7)
O	Optional
OA	Outgoing Access (CUG SS)
	Origin Address
O&M	Operations & Maintenance
OACSU	Off Air Call Set Up
OCB	Outgoing Calls Barred within the CUG
OCS	Online Charging System
OD	Optional for operators to implement for their aim
OFCS	Offline Charging System
OLR	Overall Loudness Rating
OMC	Operations & Maintenance Center
OML	Operations and Maintenance Link
OPC	Originating Point Code
OR	Optimal Routing
OS	Operating System
OSI	Open System Interconnection
OSI RM	OSI Reference Model
OSSS	Originating SMS Supplementary Service
OVI	Operators Vitally Important, such as public utilities, health, etc. which have an obligation of securing their telecommunications.
P-CSCF	Proxy Call Session Control Function
PABX	Private Automatic Branch eXchange
PAD	Packet Assembly/Disassembly facility
PCC	Policy Charging and Control
PCEF	Policy and Charging Enforcement Function. Equipment on the Gi interface which is able to analyze (Deep Packet Inspection) the IP messages and enforces the rule that it has obtained from an associated PCRF
PCH	Paging CHannel
PCM	Pulse Code Modulation
PCRF	Policy and Charging Control Function. Equipment which registers the IP handling rules for specific customers and

	pass the rule to the PCEF. PCRF is like the SCP and the PCEF is like the SSP in a CAMEL architecture.
PD	Protocol Discriminator
	Public Data
PDN	Public Data Networks
PGW	PDN Gateway (Packet Data Network Gateway), equivalent in 4G (LTE) of the GGSN. Uses GTP V2 (as latest 3G GGSN) or IPMI. Acts as an "anchor" between LTE and non LTE (e.g. WiFi) access technology.
PH	Packet Handler
	PHysical (layer)
PHI	Packet Handler Interface
PI	Presentation Indicator
PICS	Protocol Implementation Conformance Statement
PIN	Personal Identification Number
PIXT	Protocol Implementation eXtra information for Testing
PLMN	Public Lands Mobile Network
PMIP	Proxy Mobile IP, alternative non roaming simpler protocol to LTE GTPv2 over the S5/S8 interface between a SGW and a PGW allowing a single tunnel.
PNE	Présentation des Normes Européennes
POI	Point Of Interconnection (with PSTN)
PoR	Proof of Receipt (OTA SIM)
PP	Point-to-Point
PPE	Primative Procedure Entity
PPEP	Public Peering Exchange Point: the main routers of the Internet and GRX networks. They can be common with different IP addresses which separate the two IP networks.
PPR	Address Private Profile Register IP address, used with Location Services 3G, 4G
Pref CUG	Preferential CUG
PRN	Abbreviation for the MAP primitive Provide Roaming Number
PSL	MAP GSM service PROVIDE SUBSCRIBER LOCA-TION
PS	Domain Packet Service Domain (includes SGSN)
PSAP	Public Service Answering Points, such as emergency services 911(US) or 112(Europe)
PSPDN	Packet Switched Public Data Network
PSTN	Public Switched Telephone Network

PUCCH	Physical Uplink Control Chanel (used in LTE for the sending of the SRS by the UE)
PUCT	Price per Unit Currency Table
PW	Pass Word
QA	Q (Interface) Adapter
QAF	Q Adapter Function
QCI	QoS Class of Identifier
QoS	Quality of Service
R	Value of Reduction of the MS transmitted RF power relative to the maximum allowed output power of the highest power class of MS (A)
RA	RAndom mode request information field
RAB	Radio Access Bearer
	Random Access Burst
RAC	Routing Area Code (for GPRS coverage)
RACH	Random Access Channel
RADIUS	Protocol and Server used for Authentication-Authorisation and Accounting purposes
RAI	Routing Area Information (for GPRS coverage)
RAND	RANDom number (used for authentication)
RBER	Residual Bit Error Ratio
RCP	Remote Control Point (the SCCP function of a MSC (Alcatel term)
RCS	Rich Communication Suite. Uses extensions of the SIP protocol for chat and file transfers
RDI	Restricted Digital Information
REC	RECommendation
REJ	REJect(ion)
REL	RELease
REQ	REQuest
RF	Radio Frequency
RFC	Radio Frequency Channel
RFCH	Radio Frequency CHannel
RFN	Reduced TDMA Frame Number
RFU	Reserved for Future Use
RH	Roaming Hub
RLP	Radio Link Protocol
RLR	Receiver Loudness Rating
RMS	Root Mean Square (value)
RNC	Radio Network Controller UMTS 3G (equivalent to BSC (GSM 2G)

RNTABLE	Table of 128 integers in the hopping sequence
ROSE	Remote Operation Service Element
RP	Reply Path
RPOA	Recognized Private Operating Agency
RR	Radio Resource
RSE	Radio System Entity
RSL	Radio Signaling Link
RSZI	Regional Subscription Zone Identity
RTE	Remote Terminal Emulator
RTCP	Real Time Control Protocol
RTP	Real Time Protocol (used to carry 3G H323 visio sessions)
RTSP	Real Time Streaming Protocol (distributes a multimedia stream, e.g.TV) on an IP connection.
RXLEV	Received signal level
RXQUAL	Received Signal Quality
S/W	SoftWare
SABM	Set Asynchronous Balanced Mode
SACCH	Slow Associated Control Channel
SACCH/C4	Slow Associated Control CHannel/SDCCH/4
SACCH/C8	Slow Associated Control CHannel/SDCCH/8
SACCH/T	Slow Associated Control CHannel/Traffic channel
SACCH/TF rate	Slow Associated Control CHannel/Traffic channel Full rate
SACCH/TH	Slow Associated Control CHannel/Traffic channel Half rate
SAE	System Architecture Evolution, Architecture of the LTE network
SAI	Support of Area Identity
SAP	Service Access Point
SAPI	Service Access Point Indicator
SB	Synchronization Burst
SC	Service Center (used for SMS) Service Code
SCE	Service Creation Environment (scripting for IN or IVR or USSD)
SCP	Service Control Point. In an IN system, this is the SS7 part which implements the various service logics (pre-paid, number correction, anti-tromboning, etc..)
SCCP	Signaling Connection Control Part
SCF	Service Control Function
SCH	Synchronization Channel

SCLC	SCCP ConnectionLess Control
SCMG	SCCP Management
SCN	Sub Channel Number
SCOC	SCCP Connection Oriented Control
SCF	Service Control Function
SCP	Service Control Point
SCRC	SCCP Routing Control
SCTP	Stream Control Transmission Protocol (TCP with multi-homing)
SDCCH	Stand-alone Dedicated Control CHannel
SDL	Specification Description Language
SDP	Service Data Point. In an IN system this is the database with the subscriber records. It is interrogated by the SCP using the DIAMETER protocol.
SDP	Session Description Protocol, used with RCS
SDT	SDL Development Tool
SDU	Service Data Unit
SE	Support Entity
SGSN	Support GPRS Service Node.In GSM 2.5 G with GPRS, it has both circuit and IP interfaces , and provides the GPRS service to a visiting cellphone. It can deliver SMS-MT
SEF	Support Entity Function
SF	Status Field
SFH	Slow Frequency Hopping
SGi	Interface to Internet from/to a LTE PGW with normal IP packets, not encapsulated (equivalent to Gi for a 3G GGSN).
SGP	SIGTRAN Gateway Point (IP<->TDM conversion)
SI	Screening Indicator
	Service Interworking
	Supplementary Information
SID	SIlence Descriptor
SIGTRAN	Signal Transport (Working Group which works on SS7/IP)
SIF	Signaling Information Field
SIM	Subscriber Identity Module
SIO	Service Information Octet
SIP	Session Initiated Protocol (the VoIP protocol)
SLC	Signaling Link Code
SLPP	Subscriber LCS Privacy Profile
SLR	Send Loudness Rating
SLS	Signaling Link Selection

SLTA	Signaling Link Test Message Acknowledgment
SLTM	Signaling Link Test Message (polling between adjacent Point Codes)
SM	Short Message
SME	Short Message Entity
SMF	Service Management Function
SMIL	Synchronized Multimedia Integration Language (to code animated sequences for MMS)
SMG	Special Mobile Group
SMLC	Serving Mobile Location Center
SMS	Short Message Service
SMSC	Short Message Service Center
SMSCB	Short Message Service Cell Broadcast
SMSDPTP	SMS Delivery Point to Point
SMSDBCKW	SMS Delivery Backward
SMSDFWD	SMS Delivery Forward
SMS-SC	Short Message Service - Service Center
SMS/PP	Short Message Service/Point-to-Point
Smt	Short message terminal
SM-AL	Short Message Application Layer
SM-TL	Short Message Transfer Layer
SM-RL	Short Message Relay Layer
SM-RP	Short Message Relay Protocol
SN	Subscriber Number
SNM	Signaling Network Management
SNMP	Small Network Management Protocol (uses a MIB description of the equipments)
SNR	Serial NumbeR
SOA	Suppress Outgoing Access (CUG SS)
SoLSA	Support of Localized Service Areas
SON	Self Organizing Networks
SOR	Steering Of Roaming
SP	Service Provider
	Signaling Point
	SPare
SPC	Signaling Point Code
SPC	Suppress Preferential CUG
SPC	Semi Persistent Scheduling
SRES	Signed RESponse (authentication)
SRI	Abbreviation for the MAP primitive Send_Routing_Info

SRI_FOR_SM	Abbreviation for the MAP primitive Send Routing Info for Short Message
SRF	Service Resource Function
SRS	Sounding Reference Signals. The messages sent by an UE and received by the LMU in the U-TDOA positioning method.
SS	Supplementary Service
	System Simulator
SSC	Supplementary Service Control string
SSF	Subservice Field
	Service Switching function (the IN part of a MSC, may have its own GT)
SSN	Sub-System Number
SST	Sub-System Test (polling between between SCCP sub-systems)
SSTA	Sub-System Test Acknowledgement
SS7	Signaling System No. 7
SSP	Service Switching Point
STMR	SideTone Masking Rating
STP	Signaling Transfer Point
SU	Signal Unit
SUA	SCCP User Adaptation Layer
Supplicant	The low level software layer in a terminal which establishes the IP link through the Access Point
SVN	Software Version Number
SWP	Single Wire Protocol
T	Timer
	Transparent
	Type only
TA	Terminal Adaptor
	Timing Advance (between an MS and its serving BTS)
TAC	Type Allocation Code, code in general 6 digits (first of IMEI) of a handset model.
TAF	Terminal Adaptation Function
TAR	Toolkit Application Reference (OTA SIM)
TBR	Technical Basis for Regulation
TC	Transaction Capabilities
TCAP	Transaction Capability Application Part
TCH	Traffic CHannel
TCH/F	A full rate TCH
TCH/F2,4	A full rate data TCH ($\leq$2,4kbit/s)

TCH/F4,8	A full rate date TCH (4,8kbit/s)
TCH/F9,6	A full rate data TCH (9,6kbit/s)
TCH/FS	A full rate Speech TCH
TCH/H	A half rate TCH
TCH/H2,4	A half rate data TCH ($\square$2,4kbit/s)
TCH/H4,8	A half rate data TCH (4,8kbit/s)
TCH/HS	A half rate Speech TCH
TCI	Transceiver Control Interface
TC-TR	Technical Committee Technical Report
TDMA	Time Division Multiple Access
TE	Terminal Equipment
Tei	Terminal endpoint identifier
TFA	TransFer Allowed
TFP	TransFer Prohibited
TFT	Trafgic Flow Template
TI or TID	Transaction Identifier (in the TCAP protocol)
TLV	Type, Length and Value
TMN	Telecommunications Management Network
TMSI	Temporary Mobile Subscriber Identity
TN	Timeslot Number
TOA	Time of Arrival
TOE	Target of Evaluation
TON	Type Of Number
TP	Transfer Protocol (in the MAP protocol)
TRX	Transceiver
TS	Time Slot
	Technical Specification
	TeleScrvice
TSC	Training Sequence Code
TSDI	Transceiver Speech & Data Interface
TTCN	Tree and Tabular Combined Notation
TUA	TCAP User Adaptation Layer
TUP	Telephone User Part (SS7)
TV	Type and Value
TXPWR	Transmit PoWeR; Tx power level in the MS_TXPWR _REQUEST and MS_TXPWR_CONF parameters
TWLAN	Trusted WLAN, secured WLAN used for non LTE access to data services from a PGW through the S2a interface
UA	User Adaptation
UDI	Unrestricted Digital Information
UDT	Unit Data Message (of SCCP)

UDUB	User Determined User Busy
UE	User Equipment , equivalent to MS in 4G
UI	Unnumbered Information (Frame)
UIC	Union Internationale des Chemins de Fer
UL	Abbreviation for UPDATE LOCATION
UMA	Unlicensed Mobile Access
UNICAST	"UNIque broadCASTing": TV broadcast method where each reception has a dedicated channel (can be used for "on demand") and the content is "streamed".
Unicast	Shared Key between Supplicant SW in the terminal and the Acces Point to secure the Radio channel
UPCMI	Uniform PCM Interface (13-bit)
UPD	Up to date
USSD	Unstructured Supplementary Service Data
UUS	User-to-User Signaling supplementary service
UTRAN	UMTS Terrestrial Radio Acces Network (3G) equivalent to BSS (2G)
V-PCRF	Implementation of the PCRF in the VLPMN
VAD	Voice Activity Detection
VAP	Videotex Access Point
VBS	Voice Broadcast Service
VGCS	Voice Group Call Service
VLR	Visitor Location Register
VMS	Voice Mail System
VMSC	Visited MSC, (recommendation not to be used)
VPLMN	Visited PLMN
VPN	Virtual Private Network
VoIP	Voice Over IP protocol
VSC	Videotex Service Center
V(SD)	Send state variable
VTX host	The components dedicated to Videotex service
WAN	Wide Area Network
WAP	Wireless Application Protocol
WBXML	Wireless "Binary" XML
WML	Wireless Markup Language
WLL	Wireless Local Loop
WLNP	Wireless Local Number Portability
Wm	Interface in LTE between the ePDG and the AAA server
WS	Work Station
WPA	Wrong Password Attempts (counter)
XID	eXchange Identifier

XML	Extensible Markup Language
XRES	Expected Response (from the USIM card when computing A3)
ZC	Zone Code

Index

A

Access-Request, 68, 104, 110, 114
Accounting Request, 107
anti-spoof, 205
Anti-spoofing, 205
anti-steering, 7
APN change, 67
APN Correction, 96
ASN, 25, 28, 30, 61, 62, 264

B

Bellmann, 248, 249, 261
BEREC, 5, 52, 65, 66, 67, 68, 265
BGP, 2, 23, 24, 28, 30, 45, 46, 61, 265
Blind OTA provisioning, 34
Border Gateway, 28, 30, 45, 50, 59, 264, 265
broadcast, 274, 277, 286

C

CDMA, 1, 3, 4, 6, 16, 168, 169, 170, 183, 185, 186, 187, 188, 190, 192, 193, 194, 196, 198, 200, 201, 202, 223, 265, 266, 267, 274
chain of RADIUS, 110
CSFB, 150

D

device management, 30
Diameter Hub, 1, 3, 55, 59, 123
Disconnect-Request, 109
DNS, 2, 3, 11, 19, 20, 21, 22, 49, 61, 62, 64, 66, 67, 68, 73, 74, 76, 77, 78, 80, 98, 106
drift, 6, 155, 156, 159, 160, 161

E

EAP-SIM, 6, 104, 105, 106, 107, 110, 139, 140, 141, 142, 143, 144, 150, 151, 269
Echo Request, 25
emergency service, 146, 148
eNodeB, 94, 255
eSMLC, 7
eUTRAN, 250

G

geostationary, 6, 154, 159, 161, 162
Geostationary, 155, 162
GERAN, 225, 227, 228, 242
GGSN, 5, 245, 246, 249, 250, 251, 255
Globalstar, 167, 172
GLR, 7
GMLC, 59
GRX, 2, 17, 18, 19, 20, 21, 22, 23, 24, 25, 26, 28, 29, 30, 49, 50, 59, 60, 61, 62, 84, 88, 265, 267, 270, 271, 279
GSM in the aircraft, 137, 138
GTP Hub, 1, 3, 5, 49, 50, 52, 61, 62, 63, 64, 65, 67, 68, 70, 71, 72, 73, 74, 76, 82, 87, 96, 98, 100, 114, 115

H

Handover, 56, 162, 164, 165, 166, 167, 182
HLR, 2, 5, 6, 56, 59, 250
HSS, 2, 56, 59, 250, 251

About the Author

Arnaud Henry-Labordère is a graduate engineer from Ecole Centrale de Paris (1966), Ph.D (Mathematics, 1968). He was professor (chair of Operations Research) at Ecole Nationale des Ponts et Chaussées during 27 years as well as Ecole Nationale des Mines de Paris, he is currently visiting professor at Prism-CNRS (Versailles).

His industrial career was in parallel and started at IBM research (1967) as a mathematician working on the development of the very large mainframes of the time. He was then a senior engineer at SEMA and SESA directing various projects in the area of satellite ground segments, telecom networks, OS and language compilers. He then founded 3 telecom companies: FERMA, large voice mail, text-to-speech and voice recognition systems (1983), Nilcom, a pioneer in SMS Hubbing (1998) and Halys, mobile network equipment (2003) where he is chairman and chief scientist.

He is the author of nine books (six in mathematics, three in telecoms). He has been granted more than 180 patents.